COATING THICKNESS MEASUREMENT

A Guide to Methods and Instruments for

COATING THICKNESS MEASUREMENT

by

HEINZ PLOG

(*Federal Institute for Materials Testing, Berlin*)

and

C. E. CROSBY

(*EFCO Ltd., Woking*)

ROBERT DRAPER LTD.

TEDDINGTON

1971

Printed in Great Britain
by Clare o' Molesey Ltd., Surrey

FOREWORD

THIS book sets out to provide a guide to the various methods and commercially available instruments for measuring the thickness of metal and non-metal coatings. It is based on a translation of the German text 'Schichtdicken-Messung' published by the Eugen G. Leuze Verlag, Saulgau, Württ. which has been checked and edited for the present edition by Mr. C. E. Crosby who is Product Manager (Instrumentation) of EFCO Ltd., Sheerwater, England.

Coatings will normally be thinner than the substrates to which they are applied and for the purpose of this book they are taken to range in thickness from 0·1 μm to several mm. In any particular application choice of suitable coating thickness depends on many factors such as the required appearance, wear resistance, hardness, corrosion resistance, electrical properties, established specifications and of course cost, and the need to maintain quality standards has meant that the measurement of coating thickness has received much attention in metal finishing shops and materials testing laboratories[1–3]. There is no universally applicable method of test and the choice will depend on the most diverse factors such as, for example, the nature of the coating and of the basis material including the degree of surface roughness of the latter, the accuracy of the measurement required, the cost of the measuring instrument, the time taken to carry out a measurement and above all on whether it is permissible to destroy the coating or even the whole test sample. Another consideration is whether the method chosen is able, if necessary, to yield results which will be representative of the article as a whole in view of the fact that it is extremely difficult to ensure a uniform thickness of coating over the whole of an article of complicated shape.

In the comparative information presented in this book, some indication of the price of each instrument is obviously essential since price is one of the

---

1. Keinath, G. The measurement of thickness. N.B.S. Circular 585, January, 1958.
2. Barghoorn, H. Comparative investigations of coating thickness measuring methods. *Metalloberfläche*, 1965, B.7, No. 1, 8-10; No. 2, 20-25; No. 3, 36-38.
3. van der Klis, T. *Tijdschrift voor Oppervlaktetechnieken van Metalen*, 1965, 9, No. 1 (May), 17-23.

main factors which influence the choice of any particular piece of equipment. At the same time it is one of the most difficult pieces of information to include since it will vary from time to time and from place to place. It will also be influenced by modifications in design and by the number and type of accessories supplied with the equipment. Instead, therefore, of attempting to quote a firm price for each piece of equipment, a price indication is given according to the following code:

A. Under £10
B. £10 to £50
C. £50 to £200
D. Over £200.

The price indications are correct, to the best of our knowledge, in early 1971 for the basic equipment in Great Britain or the country of manufacture but obviously we cannot be held responsible for any changes or inaccuracies.

It is inevitable that the design and availability of commercial thickness testers are subject to frequent and unannounced changes. Where information has reached us in time, such changes are recorded in the text as Stop Press footnotes. Photographs and printing blocks take a little longer to produce, however, and it is inevitable that the illustrations in this book are not always of the latest designs, though this should not greatly affect the value of the book to the reader.

A summary of relevant standard specifications current in the more important industrial countries is also included together with tables of chemicals and processes for the chemical and electrolytic solution of metal coatings.

It is hoped that this extensive review will prove of assistance to all those in electroplating, galvanising, tinning, paint finishing and plastic coating shops and also to those of their customers whose sphere of interest includes the measurement of coating thickness.

January, 1971

ROBERT DRAPER LTD.,
TEDDINGTON.

PREFACE

by Prof. J. ELZE

AMONG the properties specified for metallic and non-metallic coatings, thickness is of particular importance. Although not invariably so, the thickness of a coating is nevertheless very often a decisive characteristic of the quality of a coating or of the finished article. This fact is born out in the relevant standard specifications of all industrialised countries. The question of coating thickness is a subject which frequently gives rise to considerable discussion in organisations responsible for the drawing up of standard specifications, not the least reason for this being the economic factors which are also involved.

Agreement regarding a particular coating thickness is however only meaningful if such thickness can be determined with sufficient accuracy. Development of methods of measuring coating thickness have lead in recent years to the appearance of a large number of methods and techniques. The chief objectives of this development work are to provide means of measuring the thickness of all types of coatings, to improve the accuracy of measurement to within the required tolerance limits, and to enable non-destructive testing of the finished articles. Additional requirements encountered in specific cases include the measurement of very thin coatings and measurements on very small or curved surfaces.

It can now be claimed that most of these requirements can be met. This claim may, however, have to be backed, in certain circumstances, by the availability of quite a considerable amount of equipment. A single test method and a single piece of apparatus are by no means sufficient for a satisfactory solution of all the problems that may be encountered. The practical electroplater will, therefore, have to decide in each particular case what method or what apparatus is best suited for the tests he has to carry out. His choice will depend, among other factors, on the accuracy of measurement required in order to avoid purchasing unnecessarily accurate—and expensive—apparatus. Another consideration may be whether the apparatus requires a mains supply of electricity or whether it operates on batteries and can thus be used wherever it may be needed.

In view of the very large variety of possible test methods and of apparatus on offer, the present extensive survey of the field is particularly welcome. This survey should greatly simplify the choice of test method and apparatus for the practical electroplater. Besides this it constitutes a very welcome general review of the present position of techniques for measuring coating thickness.

Prof. Dr.-Ing. Johannes Elze

CONTENTS

page

Foreword v

Preface by Prof. J. Elze viii

PART 1. DESTRUCTIVE METHODS: PRINCIPLES AND TYPICAL EQUIPMENTS

Sectioning Methods 1
Wet and/or Dry Film Meters or Gauges 2
Mesle Chord Method 7
Dial (clock) Gauge Methods 7
Chemical Stripping Methods 11
Chemical Dissolution of the Basis Metal 12
Jet and Drop Methods 13
Immersion and Spot Tests 15
Calorimetric Method 16
Electrolytic Stripping (Coulometric) Methods 17
Measurement of Breakdown Voltage 23
Spectrographic Methods 24

PART 2. NON-DESTRUCTIVE METHODS: PRINCIPLES AND TYPICAL EQUIPMENTS

Dial (Clock) Gauge and Weighing Methods 27
Magnetic Attraction Methods 27
Magnetic Induction Methods 32
Eddy Current Methods 42
Capacitative Methods 49
Thermoelectric Methods 50
Optical Methods 51
Photoelectric Method 55
Radiation Methods 56
Crystal Oscillation 68
Electrical Resistance 68

PART 3. TABLES

Table 1. Chemical Solution of Coatings: Solutions and Operating Conditions 70
Table 2. Electrolytic Solution of Coatings: Solutions and Operating Conditions 76
Table 3. Guide to Destructive Tests 84
Table 4. Destructive Thickness Measuring Instruments ... 88
Table 5. Guide to Non-Destructive Tests 92
Table 6. Non-destructive Thickness Measuring Instruments 96
Table 7. Standard Specifications 110
Table 8. Names and Addresses 118

Index 121

PART 1

DESTRUCTIVE METHODS:

PRINCIPLES & TYPICAL EQUIPMENTS

IN GENERAL, destructive methods of measuring coating thickness have the advantages that they require only relatively simple apparatus and techniques, so that determinations may be made with adequate precision in the smallest finishing shops with largely untrained personnel. The disadvantage is of course that at least part of the coating is destroyed. Tests can therefore be carried out only on a random sample basis. Additionally, measurements are usually time-consuming as, for example, when using chemical or electrolytic solution methods.

Among the destructive methods two main groups can be distinguished, namely, mechanical and solution methods, although no clear cut separation can be made between them.

Sectioning methods

The best known and possibly the most disputed destructive method, though still recommended for arbitration purposes, is that of measuring a cross section using a reflection type microscope in accordance with the procedure specified in ASTM Specification A219-58 and DIN 50 950 (April 1968). The method is, however, subject to numerous drawbacks.

Apart from the trouble involved in preparing a polished section, the method almost always necessitates the application of a backing coating[4]. This backing coating, which should have a thickness of at least 10 μm (0·0004″), is intended

4. Elze, J. and Pantzke, D. Metallographic sections of electrolytically deposited metal coatings. *Metall,* 1964, **18**, 600-603.

to prevent the rounding off of the edges during grinding and polishing and at the same time to improve the ease with which the coating to be measured can be recognised. For the latter reason it is usually also necessary to etch the section using an etchant selected to suit the particular material under examination. The finished section should deviate as little as possible from the plane vertical to the surface of the specimen.

The 1968 DIN specification differs from earlier documents in the section relating to errors of measurement. This section contains details regarding the standard deviation, the uncertainty range and the consequent relative error. According to this specification the relative error at a given point of the coating, variations in the coating thickness over the whole of the surface area not being taken into account, for an uncertainty range of $\pm$ 0·6 μm over n = 10 individual measurements of the coating thickness and for a statistical certainty of P = 95%, amounts to $\pm$ 0·8% for a coating thickness of 80 μm, $\pm$ 3% for a coating thickness of 20 μm, $\pm$ 6% for a thickness of 10 μm, $\pm$ 15% for a thickness of 5 μm, $\pm$ 30% for a thickness of 2 μm and to as much as $\pm$ 60% for a thickness of 1 μm. The error for the thinner coatings is thus hardly acceptable in a method to be used for arbitration purposes and it seems logical that other more accurate methods should be used.

Mention must also be made of the inclined section method. It differs from the conventional cross sectioning method in that the section is made at an angle to the surface. This calls for a specially prepared grinding block. The slope of the section results in a spreading of the surface to be measured and as this spreading is a function of the angle of inclination of the section, this necessitates very careful preparation of the grinding block. The accuracy which can be achieved is however less than that of the cross sectioning method as only a small microscope magnification can be used.

Wet and/or dry film meters or gauges

The measurement of the thickness of wet or dry paint coatings is carried out using mechanical devices known as wet and/or dry film meters or gauges. The wet film thickness gauges are used chiefly for checking freshly applied coatings[5, 6].

5. Hadert, H. Thickness measuring instruments. *Schleif und Poliertechnik,* 1962, No. 6, 288-290.
6. Ruder, R. Coating thickness determinations of liquid and pasty materials on substrates. *Fachber. für Oberflächent.,* 1964, April/June, 50-53.

Wet coatings

The wet coating thickness gauge developed by Henry A. Gardner and A. M. Erichson and specified in ASTM D 1212-54 (see Fig. 1) uses an eccentrically mounted measuring wheel which is attached to two concentric running wheels. Rolling of the device results in the measuring wheel dipping into the coating, the depth of immersion indicating directly the thickness of the coating; the latter can be read off on a scale provided on one of the running wheels. The range of measurement depends on the eccentricity of the measuring wheel, the upper limit being about 1500 μm (0·060 inches).

Fig. 1. Wet coating thickness gauge.

For further details see Table 4, p. 90, No. 19.

The Rossmann wet film thickness gauge (Figs. 2a and 2b) has a number of teeth. The two end teeth of the comb-like measuring element define the plane of application and penetrate through the wet film to the basis surface. The tips of the other teeth are arranged at increasing distances from the plane of application and consequently from the surface of the basis material. Some of the teeth will consequently dip into the coating while others will not touch it. The tooth which is just wetted by the coating indicates the thickness of the wet coating.

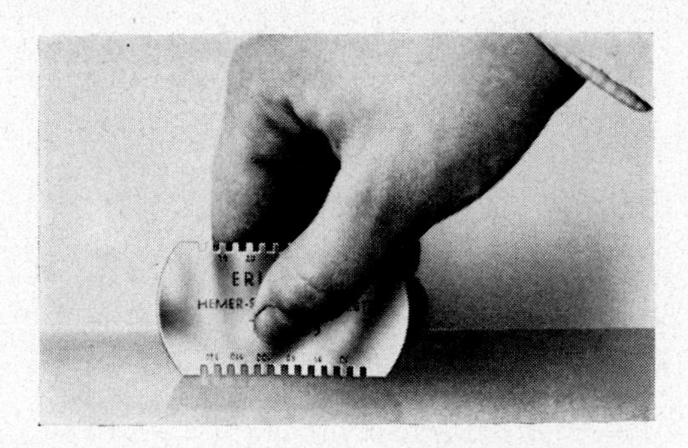

Fig. 2a. Rossmann wet film thickness gauge.

For further details see Table 4, p. 90, No. 16.

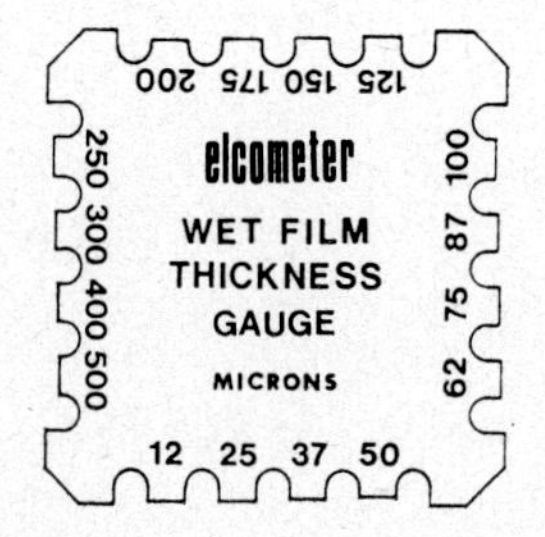

Fig. 2b. Another form of wet film thickness gauge.

For further details see Table4, p. 90, No. 16.

The Pfund film thickness meter (Fig. 3) made by the Koehler Instrument Co. consists of a convex lens the bottom surface of which has a radius of 250 mm. This convex lens forms the bottom closure of a short hollow cylinder which is arranged for free telescopic movement up and down in another hollow cylinder. A spring holds the system comprising the inner hollow cylinder and the convex lens in an initial position of rest. To carry out a measurement, the convex lens is pressed through the wet coating into contact with the basis material. After withdrawing the convex lens the diameter of the circle of immersion can be measured and the corresponding coating thickness derived from a conversion chart (or table).

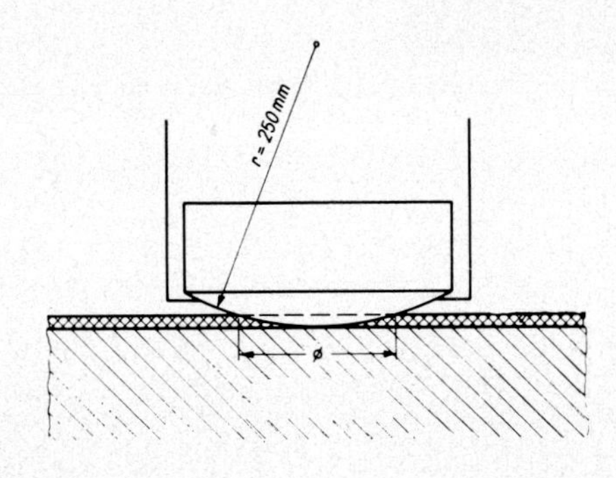

Fig. 3. Principle of the Pfund film thickness meter for measurements on wet coatings.

For further details see Table 4, p. 90, No. 13.

Dry coatings

Dry film meters are used to measure the thickness of finished, already dried coatings and the like. These instruments also depend on the penetration of the coating to the basis material by a feeler or blade and can therefore only be used for measurements on coatings the hardness of which does not exceed a certain value.

The Gardner Penetrometer (Fig. 4) is an instrument for measuring the thickness of non-conducting coatings or films on any metal. The instrument consists essentially of a needle which is driven by a micrometer screw to penetrate the coating. The pressure required for this can be varied within the range of 0-4·5 kg depending on the material of the coating. This is particularly important when measuring the thickness of soft coatings. The end of the measuring operation may be indicated by the lighting up of a signal lamp. If the zero position has previously been established by making use of a flat metal or glass surface, the thickness of the coating can then be read directly from the scale on the micrometer screw.

Fig. 4. Penetrometer dry coating thickness meter.

For further details see Table 4, p. 90, No. 12.

The Penetrometer may almost be regarded as belonging to the non-destructive group of instruments because the point size damage to the coating is in most cases negligibly small.

The PEG Thickness Gage (Fig. 5) is also manufactured by the Gardner Laboratories Inc. and is an instrument for measuring the thickness of dry coatings using a different principle. The gauge consists of two wheels mounted on an aluminium block. The smaller wheel has a diameter of about 19 mm and incorporates a cutting blade. It is rigidly fixed to the aluminium block which also serves as a handle for the gauge, while the cutting blade is located in a shallow recess on the bottom portion of the wheel and has a width of 1·5 mm. The second wheel is mounted eccentrically behind the smaller wheel. This so-called measuring wheel can be locked in position by a nut. It has a diameter of about 34 mm and carries a scale, the tangent to the two wheels forming the plane of measurement. A rotation of the eccentric measuring wheel brings about a slight change in the height of the blade on the small wheel. The ratio of the change in eccentricity to the change in height of the blade is 200 : 1.

Fig. 5. PEG Gage for measuring the thickness of dry coatings.

For further details see Table 4, p. 90, No. 11.

To carry out a thickness measurement, the measuring wheel is set to a scale position which approximates to the expected thickness of the coating. The gauge is then pushed, under a light pressure, over the coating to be measured in the direction of the small wheel with the cutting blade (direction indicated by the arrow in Fig. 5). In the process the blade reaches a cutting depth depending on the setting of the measuring wheel. The process is then repeated after altering the initial setting of the measuring wheel until the cutting blade just penetrates to the basis material. The coating thickness can then be read off directly from the scale on the measuring wheel.

The PEG Gage is particularly suited for use as a go-no-go gauge.

The Paint Inspection Gauge, the principle of which is illustrated in Fig. 6, introduced by Elcometer Instruments Ltd. is suitable for the measurement of paint films on metal, wood, plastic or concrete.

A tungsten carbide cutting tip accurately ground to 50° is drawn across the paint surface. The ‘V’ cut produced by the cutter is then examined by the special microscope and built in illuminator fitted to the gauge, (magnification × 50). As the angle of the cut is 45°, the width of coating is the same as the depth so that a direct reading of the film thickness on the various substrates can be made without any calculations being necessary. The maximum thickness range is 2 mm and with special cutting tips, accuracies of ± 0·5 microns are possible.

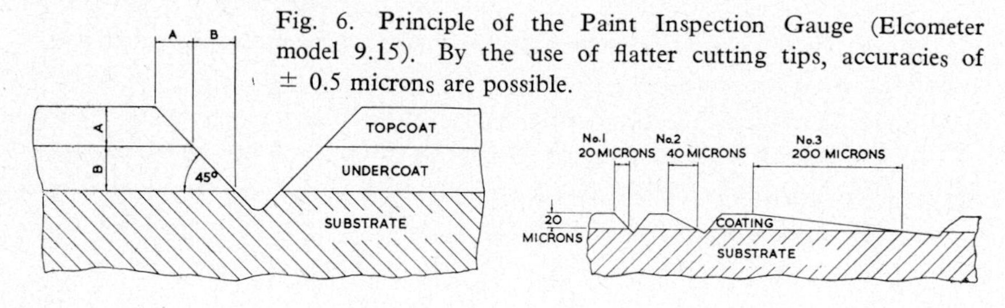

Fig. 6. Principle of the Paint Inspection Gauge (Elcometer model 9.15). By the use of flatter cutting tips, accuracies of ± 0.5 microns are possible.

Mesle chord method

A method developed by Mesle which is mainly applicable to metal coatings requires only the use of a face grinding wheel for measurements on flat surfaces or of a file for measurements on cylindrical surfaces. This is the Mesle chord or arc section method (Fig. 7) in which a fine grain face grinding wheel is used to grind through the coating to the basis material. Having measured the diameter D of the grinding wheel and the length S of the arc, the thickness of the coating d can be calculated from the expression $d = \frac{S^2}{4D}$. For measurements on cylindrical surfaces the appropriate values are substituted in the expression. The possible difficulty of detecting the end point of the grinding process may be avoided by continuing to grind into the basis material. In such a case the depth of penetration d_1 must be subtracted from the calculated value of the coating thickness.

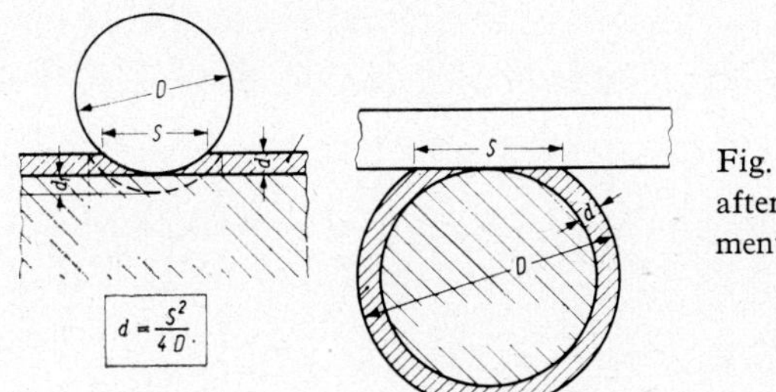

Fig. 7. Arc section method, after Mesle, mostly for measurements on metal coatings.

If the transition from the coating to the basis material cannot be clearly distinguished, e.g. because both materials have the same colour, treatment of the surface with chemical reagents capable of bringing about a colour change will be necessary. The accuracy of the method is ± 10%.

Dial (clock) gauge methods

A more-or-less universally applicable method for solid coatings is the measurement of thickness difference using a clock or dial gauge or other suitable precision measuring instrument. This method can be used to determine the thickness of practically all existing coatings including dry paint and electro-deposits. The only exceptions are coatings the formation of which occurs wholly or partially (e.g. as in anodising) by the coating growing into the basis material. Different methods are used to measure the difference in height between the surface of the basis material and that of the coating.

One method is to carry out the measurements before and after applying the coating. This method is however restricted and can only be applied to thick coatings. Moreover, the assumption which it involves that the thickness or the diameter of the basis material remains the same within certain limits is not always justified.

A better method renders the measurement of the coating thickness independent of the thickness of the basis material. It involves masking the surface at the point at which the thickness of the coating is later to be measured. After the coating has been applied, the mask is removed and the resulting difference in height can then be measured.

Another method depends on the local removal of the coating either mechanically or electrolytically. Chemical or electrolytic solution (see pp. 11 and 17) requires a preliminary degreasing and the subsequent thorough cleaning of the area concerned. Many arrangements comprising a precision measuring instrument and a suitable stand are used as instruments for measuring coating thickness.

Examples of such instruments are the Millimess (Fig. 8) by C. Mahr and the Leitz Tolerator (Fig. 9). The Platimeter (Fig. 10), manufactured by T. Rapp, is specifically intended for measuring coating thickness. The baseplate of the stand is provided with clamping strips for holding small articles, which need not necessarily be flat. Other clock gauge instruments require flat surfaces.

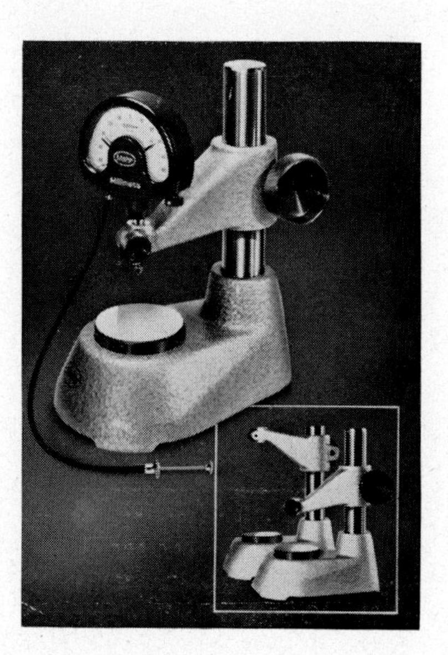

Fig. 8. Millimess precision measuring gauge and table.

For further details see Table 4, p. 90, No. 9.

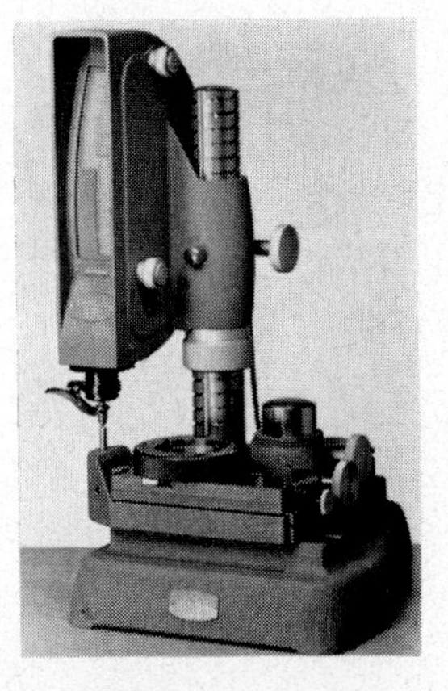

Fig. 9. Leitz Tolerator.

For further details see Table 4, p. 90, No. 18.

Fig. 10. Platimeter.

For further details see Table 4, p. 90, No. 14.

The great advantages of all these instruments are their ease of handling and their portability. They all have two fixed supports below the clock gauge and between them is the feeler point which can move vertically up and down. This feeler point serves to transmit the difference in height between the surface of the coating and an exposed portion of the base to the pointer of the gauge. A glass plate can be used for the initial setting of the gauge. Instruments of this type include the Coating Thickness Meter type S 1566 (Fig. 11) by H. C. Kröplin, the Rossmann Dry Film Thickness Gauge type 233 (Fig. 12) which meets the requirements of ASTM D 1005-51 and is manufactured by A. M. Erichsen, being sometimes referred to as the IG-Watch, and the Rossmann Wet and Dry Film Thickness Gauge type 296 (Fig. 13).

Fig. 11. Coating thickness meter type S 1566.

For further details see Table 4, p. 88, No. 2.

Fig. 12 Rossmann dry film thickness gauge type 233.

For further details see Table 4, p. 90, No. 15.

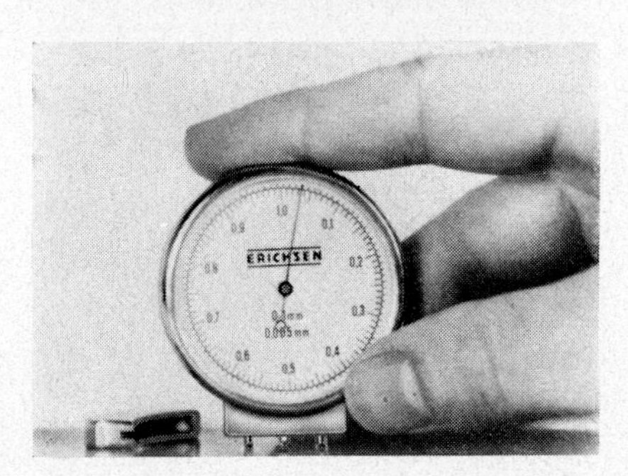

Fig. 13. Rossmann wet and dry film thickness gauge type 296.

For further details see Table 4, p. 90, No. 17.

The last named instrument is of essentially the same design as the other two instruments referred to but is in addition provided with a screw mounted on the upper portion of the case by means of which the otherwise freely moving feeler can be raised or lowered manually so as to rest on the surface of a wet film.

The Dektak Surface Profile Measuring System of Sloan Instruments Division measures and records surface line profiles by scanning the surface with a diamond stylus at approx. 15 mg tracking force.

Chemical stripping

In this method[7] the coating, which is usually metallic, is dissolved using a suitable chemical solvent. To dissolve the coating only solvents can be used which are capable of completely dissolving the coating without, or whilst only very slowly, attacking the basis metal (Table 1, p. 70). The thickness of the coating can then be obtained by measuring the difference in thickness before and after stripping either by a dial gauge method or from the difference in weight determined by weighing.

The determination from the difference in weight using the following expression gives the mean thickness of the coating:

Coating thickness in μm =

$$\frac{100 \times \text{weight of dissolved coating}}{\text{density} \times \text{area of dissolved coating}} \qquad \left\{\frac{\text{g}}{\text{g/cm}^3 \times \text{dm}^2}\right\}$$

This method of measuring coating thickness forms the subject of several standard specifications (see Table 7, p. 110). Its drawback is that in general one obtains only the mean thickness of the coating over the surface as a whole. Since the actual coating thickness may vary widely from point to point on the surface it may be necessary to carry out the determination on a clearly defined portion of the surface, the surrounding areas being masked using, for example, a suitable varnish. In such cases the thickness of the coating can be obtained by determining analytically the amount of metal dissolved, instead of using the difference in weight.

Various methods for the quantitative analytical determination of the amount of metal dissolved are available. P. Diesing and H. Schneider[8] have described a method in which titration with methylene blue (2·29 g methylene blue in 1 litre of water) is used for the quantitative determination of tin coatings on brass and copper. The test sample is placed in boiling (1:1) hydrochloric acid and the methylene blue solution is added continuously until a constant blue colour denotes the end point of the titration. The mean coating thickness is then obtained from the number of ml of methylene blue solution used per cm^2 of the surface.

The thickness of a zinc coating on wire can be obtained in g/m^2 using the gas-volumetric method specified in DIN 51 213 (July 1965). In the so-called Zaba apparatus (Fig. 14), which was specially developed by Keller and Bohacek for use in this method, wires up to 11 mm in diameter can be tested. A wire specimen of known length is introduced into the apparatus which is

7. Mathur, P. B. and Lakshmanan, A.S. Chemical methods of testing the thickness of electrodeposits. *Electroplating and Metal Finishing*, 1962, **15**, No. 4, 114-119.
8. Diesing, P. and Schneider, H. Two new laboratory methods for testing electrolytic deposits. *Galvanot. u. Oberflächenschutz*, 1963, **4**, No. 9, 203-211.

filled with acid. The gas formed is collected in a glass cylinder and its volume read off on a scale engraved on the cylinder. Knowing the test temperature and the diameter of the wire the weight of the zinc coating in g/m^2 can be obtained from the measured volume of gas with the aid of tables supplied with the apparatus. By carefully observing the operating instructions measurements with an accuracy of $\pm$ 1 to 2% can be achieved. A special version of the Zaba apparatus is available for determining the weight of zinc coatings on narrow strip.

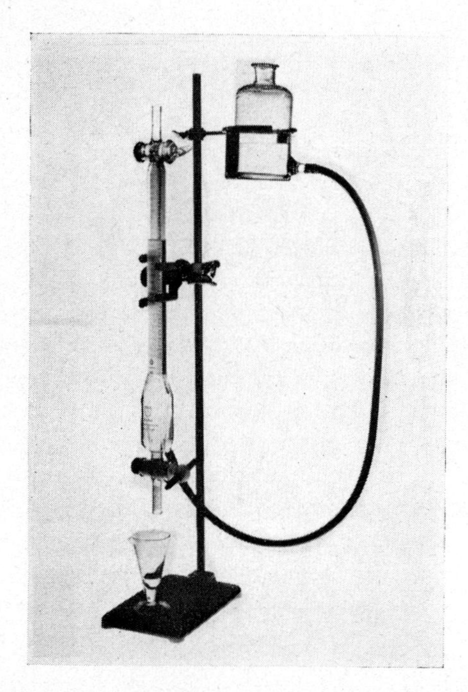

Fig. 14. ZABA apparatus for determining zinc coatings on wire by the gas volumetric method.

For further details see Table 4, p. 90, No. 20.

Chemical dissolution of the basis metal

A method which is only applicable to coatings of noble metals such as platinum and gold involves the chemical dissolution of the basis metal. A very precise calculation of the test surface is required and the specimen is then immersed in a suitable acid until the basis metal has been dissolved completely away. The undissolved coating of the noble metal is rinsed, dried and then weighed on an analytical balance. The mean thickness of the coating can then be calculated while the thickness of the coating at any given point (local thickness) can be obtained by direct measurement with a micrometer or other precision measuring instrument.

The accuracy of this method depends to a large extent on the accuracy of the preliminary calculation of the surface area.

Jet and drop methods

The jet method developed by the British Non-Ferrous Metals Research Association in England has been standardised in Germany, the U.S.A. and Great Britain where it is known as the BNF Jet Test[9-11]. It permits the measurement of the thickness of electrodeposited coatings with an error of $\pm$ 15%, a stop watch being used to measure the time taken by a specified jet of penetrating solution to dissolve the deposit at the point of impingement of the jet. The end point of the penetration is reached when the basis metal becomes visible under the jet. It is found in practice that where the deposit and the basis metal have the same appearance, the recognition of the end point can be very difficult in spite of the use of suitable indicator solutions. The thickness of the deposit is calculated from the time required to penetrate the deposit multiplied by the rate of removal of deposit, the rate of removal depending on the material and being very much influenced by temperature. The rate of removal can be obtained from curves or tables given in the booklet supplied with the apparatus and reproduced in DIN 50 951 and elsewhere. The method is not entirely satisfactory but although attempts are being made to exclude it from standard specifications, it will nevertheless still remain of some importance, at least in surface finishing and related fields because of the ease with which measurements on curved surfaces can be carried out.

Fig. 15 is a diagram of the apparatus while a typical rate-of-removal curve is shown in Fig. 16. The penetrating solutions employed are as follows:

For nickel, cobalt, copper, bronze and composite nickel-copper-nickel coatings:

Ferric chloride ($FeCl_3.6H_2O$)	300 g
Cupric sulphate ($CuSO_4.5H_2O$)	100 g
Distilled water to	1 litre

9. Clark, S. G. *Ann. Proc. Amer. Electroplaters' Soc.*, 1939, **27**, 24-30.
10. Read, H. J. and Lorenz, F. R. Comparison of methods for nickel on steel. *Plating*, 1951, **38**, 255-263.
11. Read, H. J. and Lorenz, F. R. Comparison of methods for acid copper on steel. *Plating*, 1951, **38**, 945-952, 958.

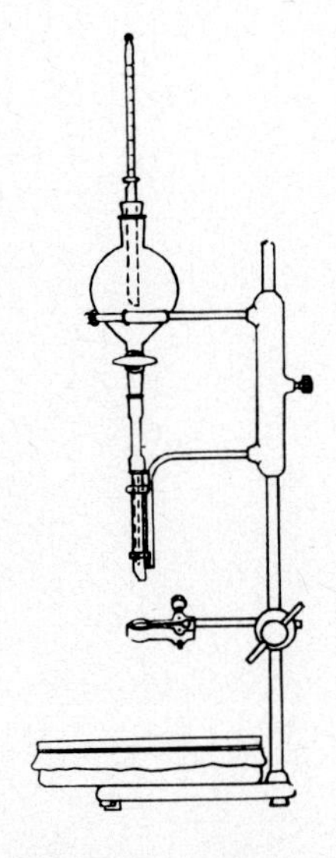

Fig. 15. B.N.F. Jet Test apparatus.

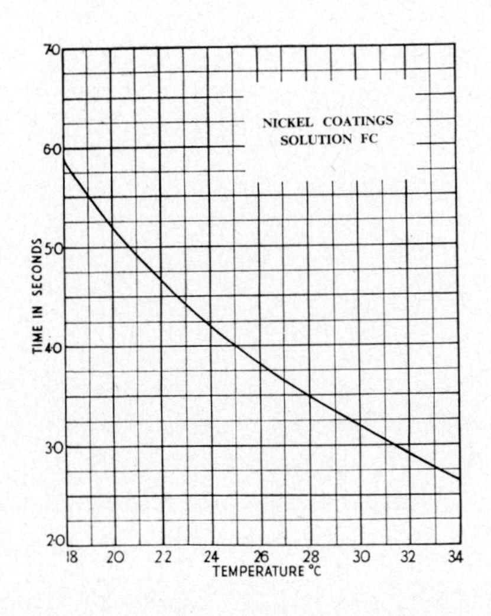

Fig. 16. Rate of penetration of nickel coatings by ferric chloride solution in the Jet Test.

For cadmium coatings:	
Ammonium nitrate	17·5 g
N/1 Hydrochloric acid	15·5 ml
Distilled water to	1 litre

For zinc coatings:	
Ammonium nitrate	70 g
N/1 Hydrochloric acid	70 ml
Distilled water to	1 litre

For silver coatings:	
Potassium iodide	260 g
Resublimed iodine	7.44 g
Distilled water to	1 litre

For tin coatings:	
Trichloro-acetic acid	100 g
Distilled water to	1 litre

For tin-zinc coatings:	
Trichloro-acetic acid	50 g
Distilled water to	1 litre

For lead coatings:

Acetic acid (glacial)	1 vol.
Hydrogen peroxide (20 vol.)	1 vol.
Distilled water	3 vol.

For chromium coatings:

Antimony oxide (Sb_2O_3)	20 g
Hydrochloric acid (sp. gr. 1·16) to	1 litre

The drop test, which in the U.S.A. forms the subject of ASTM A 219-58, operates in a similar manner to the jet test except that the penetrating solution impinges on the test surface not in the form of a jet but as a steady succession of drops falling at the rate of 100 $\pm$ 5 drops per minute.

Immersion and Spot Tests

The immersion method developed by Preece is one of the oldest chemical solution methods. The test sample is immersed in the solvent for certain intervals of time. The number of immersions required to dissolve off the coating is a measure of the coating thickness. The results obtained by this method are not very reliable.

The spot test[12] according to BS 1223-1959, ASTM A219-58 and DIN 50 953-1968 was developed for the purpose of determining the thickness of very thin chromium deposits on nickel, copper-nickel or steel undercoatings. A wax ring is drawn round a portion of the surface, e.g. using a Chinagraph pencil or the like, and within this ring a drop of 11.5 N hydrochloric acid is applied to the surface. The time from the start to the finish of gas evolution or up to the appearance of the yellowish nickel undercoating is measured with a stop watch. The time taken to penetrate the coating multiplied by the rate of dissolution, the latter being taken from graphs reproduced in the specifications, gives the coating thickness in μm.

According to comparative tests carried out at the Bundesanstalt für Materialprüfung in Berlin, this standardised test method is applicable only to coatings have a thickness of up to about 0·6 μm (see Fig. 17), and this fact has been taken into consideration in the latest edition of the DIN specification published in 1968. The test is unsuitable for thicker deposits because the end point of the dissolution of the chromium deposit becomes no longer readily

12. Jogarao, A., Sundaram, M. and Subramanian, R. Spot test for measuring the thickness of brass platings. *J. Sci. Industr. Res.*, 1962, **21 D**, 132-133.

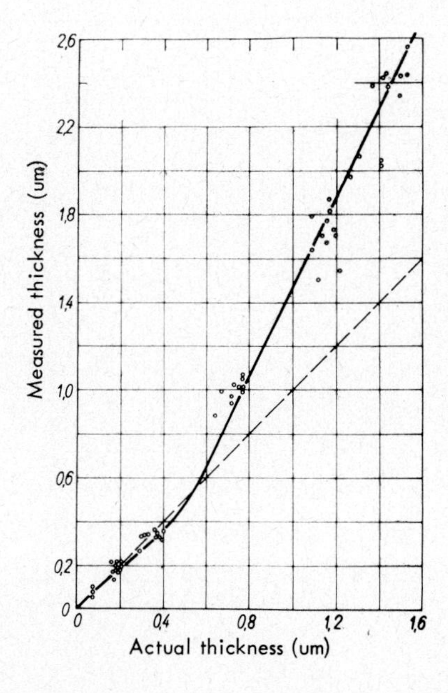

Fig. 17. Comparison of test results obtained for the thickness of chromium on nickel using the spot test and the coulometric method.

——— Spot test.
- - - - - Coulometric method.

observable due to the vigorous evolution of gas and the green colour which develops. In such cases the values obtained are considerably higher than the true thickness by an amount of the order of + 50%. The test method, which is very simple to carry out, can be used successfully in plating practice where the chromium deposits have a thickness below 0·6 μm.

Calorimetric method

G. Krijl and J. L. Melse[13] have described a method by means of which the mean coating thickness can be determined from the heat of reaction of metals in organic solvents. A simplified design of calorimeter is used to determine the relation between the thickness of the coating and the amount of heat evolved. The method has been used with nickel, zinc and cadmium coatings on steel and with silver coatings on phosphor bronze. Each metal requires a suitable solvent. The amount of heat evolved is proportional to the thickness of the coating.

13. Krijl, G. and Melse, J. L. Calorimetric determination of metal coating thicknesses on small objects. *Trans. Inst. Met. Fin.*, 1961, **38**, Pt. 1, 22-26.

Electrolytic stripping (Coulometric) methods

In electrolytic stripping methods[14–16] the metal coating is made the anode, in contrast to electrodeposition. The method is preferred where rapid and definite stripping of the coating is required. Additionally it is often possible to use an electrolyte which will not attack the basis metal at all. An example of this is provided by stripping in cyanide solutions in cases where the basis metal is iron.

The basis of the electrolytic stripping method is the fact (Faraday's law) that dissolution of 1 gramme equivalent of a metal always requires the same amount of electricity ($I \times t$), namely 96 494 C (coulombs). As 1 A = 1 C/s, stripping of 1 gramme equivalent of metal in 1 hour requires

$$\frac{96490}{3600} = 26{\cdot}8 \text{ A}$$

The weight G of metal dissolved can then be obtained knowing the current efficiency η, which in the ideal case should be equal to 100%:

$G = c \times I \times t \times \eta$ where

G = weight in g

I = current in A

t = time in h

$$c = \text{electrochemical equivalent} = \frac{\text{atomic weight}}{\text{valency} \times 26{\cdot}8} \left\{\frac{\text{g}}{\text{Ah}}\right\} \text{ and}$$

$$\eta = \text{current efficiency} = \frac{\text{amount of metal actually deposited}}{\text{amount theoretically possible to deposit}}$$

From the density

$$\rho = \frac{G}{V} \text{ in g/cm}^3$$

of the metal dissolved and making use of the relation $d' = V/F$ (V = volume in cm^3, F = surface area of coating dissolved in cm^2, d' = thickness of coating in cm), the thickness d in μm of the coating is obtained from the expression:

$$d = \frac{G}{\rho \times F} \times 10^4.$$

14. Baier, S. W. The B.N.F. Plating Gauge. *Confidence in Plating*, Feb. 1960, S.D. 2/D 10.
15. Meuthen, B. The coulometric determination of coating thickness of metallic deposits on steel strip, plate and tubes. *Bänder, Bleche, Rohre,* 1965, **6**, No. 9, 513-519.
16. White, R. A. Coulometric plating thickness meter. *Metal Industry,* 1961, **98**, No. 23, 455-458.

The measurement of the thickness of electrodeposits by this method is described in Standard Specification DIN 50 955.

For coatings of tin and zinc on steel it is possible, by recording the potential-time diagram[17], to determine, in addition to the thickness of the coating, the process used for applying the coating, e.g. hot dip galvanising, etc. For the test, a specimen of known surface area is placed in an electrolytic cell and made the anode in a d.c. circuit. The cathode can be made of a sheet of any metal which does not dissolve in the electrolyte. The potential measurement is carried out, preferably without the flow of current, using a reference electrode (e.g. a calomel or Ag/AgCl electrode) via a capillary tube located near the surface from which the coating is being removed. If no absolute values are required a measurement of the cell voltage will be sufficient.

It is preferable to use a recording instrument for measuring the potential, such an instrument serving at the same time to determine the time required to strip the coating. The potential-time diagrams of a hot dipped zinc coating and of an electrodeposited zinc coating are shown in Fig. 18. This process is standardised in DIN 50 932 (1971).

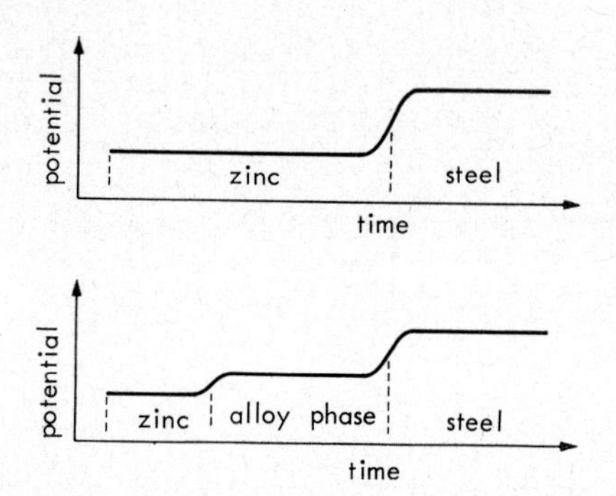

Fig. 18. Anodic dissolution curves for zinc coatings on steel. Upper diagram: electrodeposited zinc. Lower diagram: hot dipped zinc coating.

Fig. 19 shows an electrolytic cell developed for use in this process by W. Katz and H. Gehrke in the Bundesanstalt für Materialprüfung. The cell is made of transparent plastics. Two permanent magnets are used to press a sealing ring onto the test surface to define the area from which the coating is to be stripped; this arrangement enables the cell to be used for measurements on inclined or vertical surfaces. Where the basis metal is steel the thickness of other conducting coatings can be measured. If the basis metal is not very magnetic and provided its thickness is not too great, a ferromagnetic backing material can be placed on the opposite side in order to ensure that the magnets are able to hold the cell sufficiently firmly against the surface.

17. Hoare, W. E. and Britton, S. C. Tin plate testing. Tin Research Institute, 1960.

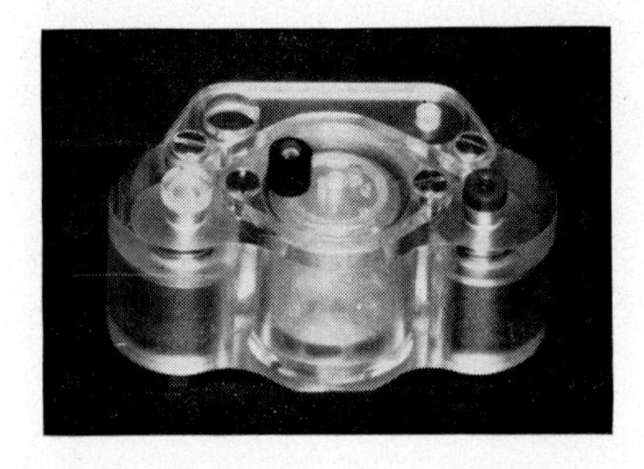

Fig. 19. Test cell for anodic stripping (after Katz and Gehrke).

The Seddon test[18] was developed for determining the thickness of tin coatings on copper wire. In this test the tinned copper wire is placed at the centre of a hollow cylindrical cathode of copper sheet. The electrolyte consists of a solution of 25 g of crystalline stannous chloride in 100 ml of hydrochloric acid made up to 500 ml with distilled water. During the test a sudden change in the current flowing through the d.c. circuit indicates the complete removal of the pure tin coating. The time between this first and a second change in stripping current corresponds to the thickness of the alloy coating.

Francis has developed an apparatus of this type intended specifically for testing electrodeposited tin coatings. A relay is provided which automatically switches off the apparatus when a change in potential occurs. The time required to strip the coating is shown by a clock and the thickness of the coating can then be obtained by reference to a table. A heavy cylinder of stainlesss teel is placed on the test surface and forms the positive contact. The cathode, which is also of stainless steel, is arranged inside the cylinder and takes the form of an insulated tube which is moveable vertically and which constitutes the test cell. A 10% solution of sodium hydroxide is recommended as the electrolyte. The stainless steel tube, the lower end of which is provided with a rubber ring, is pressed against the test surface by a spring.

The same principle is used in the well known Kocour apparatus, Model 660 (Fig. 20). This is suitable for measuring the thickness of practically all common electrodeposits which are electrically conducting, including composite deposits such as copper-nickel-chromium. A special electrolyte is required for each particular combination of metals and these electrolytes are supplied with the apparatus. Various accessories enable the measuring range to be extended to below 0·01 μm, this being particularly useful for chromium deposits (Fig. 21) and also permitting measurements to be made on small articles and wires.

18. Kalpers, H. Determination of the coating thickness of tin deposits. *Metall,* 1953, **7**, No. 3-4, 120-121.

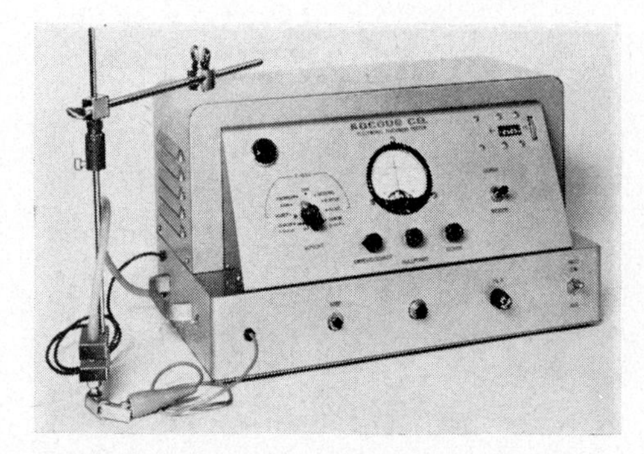

Fig. 20. Kocour coating thickness measuring apparatus, Model G-660 (coulometric).

For further details see table 4, p. 88, No. 8.

Two measuring cells of different sizes provide test areas of 2·4 and 3·2 mm diameter. The measuring cells are made of Monel metal and are quickly interchangeable. Vigorous agitation within the measuring cell is effected by introducing a small air stream. In the earlier models of this apparatus stirring of the electrolyte, which has a volume of about 2 ml, is effected by a small plastics stirring rod driven directly by a motor.

Fig. 21. Measuring attachment GMT for use with the Kocour coating thickness measuring apparatus to enable extension of the measuring range to 0.07 μm (chromium 0.007 μm).

The end point of the measurement is indicated by the lighting up of a signal lamp and the ringing of a bell. The use of the bell, which is available as an extra, is recommended for measurements on thicker coatings where the test takes correspondingly longer.

The coating thickness can be read off directly in μm. The apparatus requires a mains supply of 220 V, 50 Hz. Kocour apparatus models the designation of which does not include the letter G, which stands for German, have the counter relay scale calibration in 10^{-5} inches and are designed for a mains supply of 110 V, 60 Hz.

Instruments which closely resemble the Kocour apparatus are the Coulometric Plating Gauge Mk. III manufactured by Thorn Bendix Ltd. and the Coulometric Plating Thickness Meter manufactured by M. L. Alkan Ltd. (Fig. 22) which has become more and more popular as an accurate and rapid method for thickness testing in plating departments and laboratories. The 'practical' accuracy is rather better than 5%, i.e. as good as most microsectioning, and considerably more accurate than microsectioning for thin coatings of say 8 microns or less. Although the test is destructive in as much as it destroys a small area of the coating (but not the whole article as in microsectioning), it has the advantage of being a 'primary' method (like 'strip and weigh') and does not require to be calibrated against standards whose thickness has to be measured by some other means.

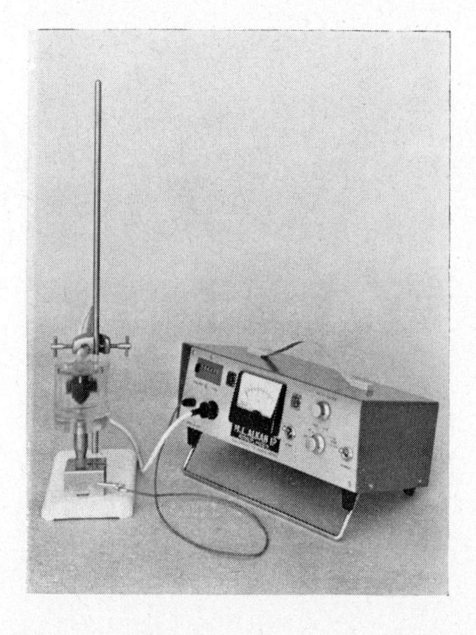

Fig. 22. Model SS Alkan Plating Thickness Meter (Coulometric principle).

For further details see Table 4, p. 88, No. 4.

Both manufacturers have recently re-designed their versions of the B.N.F. Coulometric Thickness Meter. The new Thorn Bendix instrument, known as their Mark III Meter, is supplied with two cells, a standard cell for accurate results and a small cell which can be employed in somewhat recessed areas and on quite curved surfaces or on small areas, but which is not expected to give quite as accurate results as the standard cell. The most important features of the new Alkan instrument (the Model SS) are a digital coulomb integrator for easy read out and completely solid state circuitry for maximum reliability. This new compact instrument measures 12″ × 8″ × 4″ high and weighs only 8 lb. It will measure deposits from 0·000002″ to 0·002″ with an accuracy of 2·5%.

The solutions recommended for use in the coulometric test are given in the accompanying table.

Recommended Electrolytes for Coulometric

Thickness Testing of Electrodeposited Coatings

COATING	SUBSTRATE									
	Steel		*Copper and brass*		*Nickel*		*Aluminium*		*Zinc*	
Cadmium	A1	B1	A1	B1	A1	—	A1	—	—	—
Chromium	A2	B2	A3	B3	A2	B4	A2	B4	—	—
Copper	A4	B5	—	—	A4	B6	A4	—	—	B7
Lead	—	B8	—	B8	—	B8	—	—	—	—
Nickel	A5	B9	A5	B10	—	—	A5	B9	—	—
Silver	A6	—	A7	B11	A6	—	—	—	—	—
Tin	A3	B12	A3	B12	A3	—	A2	B13	—	—
Zinc	A8	B14	A8	B14	A8	—	A8	—	—	—

A1 100 g/l KI with traces of I_2

A2 100 g/l Na_2SO_4

A3 73 g/l HCl (dilute 175 ml HCl (sp.gr. 1·18) to 1 litre)
or 150 g/l NaOH

A4 80 g $NaK_4H_4O_6$ (sodium potassium tartrate)
+ 100 g NH_4NO_3 } in 1 litre

A5 30 g NH_4NO_3
+ 30 g NaCNS } in 1 litre

A6 100 g $NaNO_3$
+ 5 g HNO_3 } in 1 litre

A7 180 g/l KCNS

A8 100 g/l NaCl

B1 30 g KCl
+ 30 g NH_4Cl } in 1 litre

B2 186 g H_3PO_4
+ 10 g CrO_3 } in 1 litre (dilute 118 ml orthophosphoric acid (sp.gr. 1·75) and 10 g CrO_3 to 1 litre)

B3 100 g/l Na_2CO_3 (for coatings up to 5 microns)

B4 98 g/l H_3PO_4 (dilute 64 ml orthophosphoric acid (sp.gr. 1·75) to 1 litre)

B5 800 g NH_4NO_3
+ 10 ml NH_4OH (sp.gr. 0·88) } in 1 litre

B6 100 g K_2SO_4
+ 20 ml H_3PO_4 (sp.gr. 1·75) } in 1 litre

B7 Pure H_2SiF_6 solution containing not less than 30% H_2SiF_6 (Slightly weaker acid may be used if some $MgSiF_6$ is added to the solution)

B8 200 g CH_3COONa + 200 g CH_3COONH_4 } in 1 litre

B9 800 g NH_4NO_3 + 3·8 g thiourea } in 1 litre—Make up fresh every 5 days

B10 42 g/l HCl (dilute 100 ml hydrochloric acid (sp.gr. 1·18) to 1 litre)

B11 100 g/l KF

B12 100 g KNO_3 + 100 g KCl } in 1 litre
Or preferably 175 ml HCl (sp.gr. 1·18) in 1 litre (= A3)

B13 90 g H_2SO_4 + 5 g KF } in 1 litre (dilute 50 ml sulphuric acid (sp.gr. 1·84) and 5 g KF to 1 litre)

B14 100 g/l KCl (or NaCl as A8)

Before testing, the area should be cleaned with a cloth wetted with an organic solvent for grease removal, if necessary. For certain coatings it may be necessary to employ an additional treatment to ensure that the surface is free from passive films or any conversion coating.

Mention must also be made of the Bendix method intended for tin coatings. Here the tin coatings are stripped anodically in hydrochloric acid containing an excess of iodine, the amount of tin then being determined by back-titrating the amount of iodine consumed. The method is particularly suited for rapid routine testing. A test on a specimen having a surface area of about 25 cm^2 using a current of 3 A takes 1-3 minutes.

Measurement of breakdown voltage

The breakdown voltage of thin insulating coatings is approximately proportional to the thickness of the coating. This fact can be used for measuring the thickness of anodic axide coatings and a simple circuit is shown in Fig. 23.

The voltage required for breakdown is obtained from a variable transformer starting from zero. A normal potential-measuring instrument can be used to

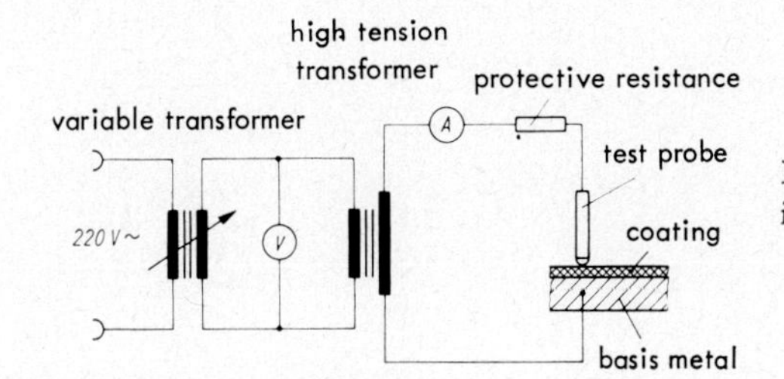

Fig. 23. Circuit for measuring breakdown voltage.

measure the potential on the input side of the high tension transformer. This reading multiplied by the step-up ratio of the high tension transformer then gives the breakdown voltage applied to the anodic coating between the basis metal and the probe. The occurence of breakdown of the anodic coating is indicated by the sudden increase in the reading of a current measuring instrument. Damage to the latter is prevented by a protective resistance of 100-500 kΩ. The probe should be applied to the test surface under a constant pressure and should be provided with a replaceable tip in order to obtain reproducible values. As damage to the test surface is restricted to a minute area this method may be regarded as almost non-destructive.

An instrument based on the breakdown voltage principle is the Gratometer (Fig. 24) developed by Langbein-Pfanhauser in Vienna. It enables measurements on coatings having a thickness of up to 25 μm for which the breakdown voltage amounts to 1,500 V. The high voltage can be adjusted continuously up until the breakdown of the anodic coating.

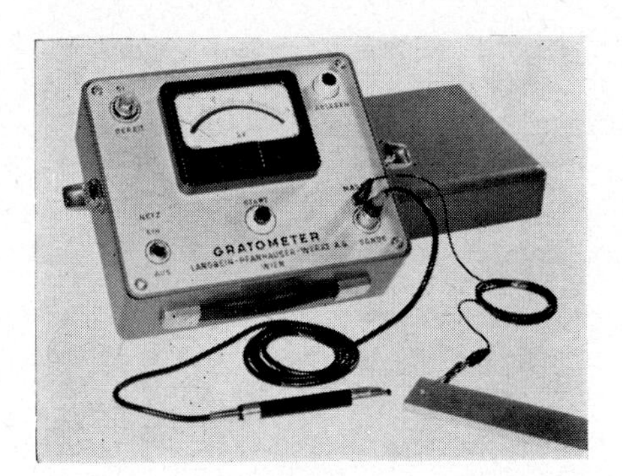

Fig. 24. The Gratometer which measures coating thickness by determining the breakdown voltage.

For further details see Table 4, p. 88, No. 6.

A similar instrument is the Laboratory Tester manufactured by Bell Telephone Laboratories Inc. (p. 88).

Spectrographic methods

Two spectrographic methods, the comparison method and the sparking method, are based on the differences in the spectra of the coating and the basis metal. The comparison method uses a spark spectrograph and depends on the reduction in the intensity of the spectrum of the basis metal caused by the presence of the coating. By comparing the spectrum of the test specimen with the spectra of specimens bearing coatings of known thickness and of exactly the same composition, an indication of the thickness of the coating on the test specimen can be obtained.

The sparking method does not necessarily require the use of reference specimens. Sparking of the specimen is continued until the spectrum of the basis metal is clearly apparent. The sparking time then provides a measure of the coating thickness.

PART 2

NON-DESTRUCTIVE METHODS: PRINCIPLES AND TYPICAL EQUIPMENTS

THE non-destructive methods have become increasingly important in recent years [19, 20]. Apart from not spoiling the part tested, the non-destructive methods also possess other advantages which include the rapid testing of large numbers of parts, the testing of small and very small parts, the possibility of measuring the thickness of coatings on the interioir of pipes and tubes above a certain diameter, and the routine control of production. The applicability of the methods is however limited by the variety of possible combinations of materials encountered in practice. There is no non-destructive method which can be used universally irrespective of the conductivity, permeability, atomic number or some other property of the material.

In considering the application of non-destructive methods it is necessary to distinguish three groups of materials (Fig. 25):

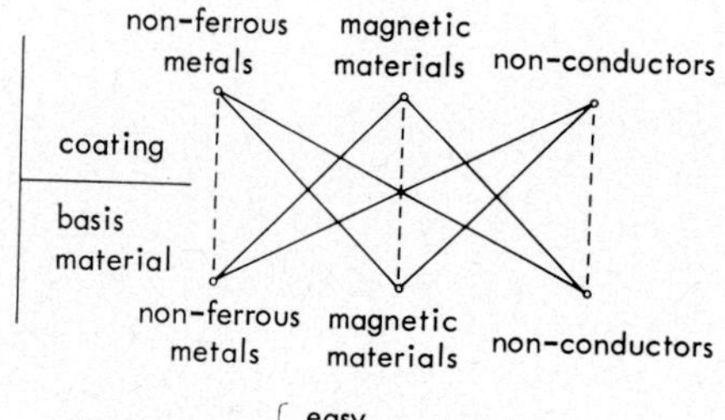

Fig. 25. Ease of measuring different combinations of coating and substrate. (After P. Müller).

---

19. Makiola, C. Coating thickness determination methods. Mitt. Forschungsges. *Blechverarbeitung,* 1963, No. 20-21, 297-303.
20. Müller, P. Methods and instruments for non-destructive coating thickness measurement. *Metall,* 1964, **18**, No. 9.

1. Ferro-magnetic materials,
2. Non-ferrous metals and
3. Electrically non-conducting materials.

This grouping applies both to the coatings and to the basis materials. Those cases where the coating and the basis material belong to two different groups, e.g. a non-ferrous metal on a ferro-magnetic base or a non-ferrous metal on a non-conducting material, are relatively easy to deal with. Those cases where the coating and the basis material belong to the same group are more difficult.

Dial gauge and weighing methods

Possibly the only exceptions to what has just been said are the methods of measuring coating thickness with the aid of a dial or clock gauge or by determining the difference in weight. These have already been dealt with among the destructive methods but in this case, for non-destructive testing, the difference in thickness or weight is obtained by measurements taken before and after the application of the coating. This naturally limits the use of the methods to the manufacturer and, generally, to the manufacturer operating his own coating process.

Using the weighing method under optimum conditions, as for example in the preparation of test reference specimens where uniformity of coating thickness, an accurate calculation of the surface area and weighing using an analytical balance are possible, an accuracy of better than $\pm$ 1% can be expected.

Magnetic attraction methods

These methods are suitable for measuring the thickness of non-ferromagnetic coatings on a ferromagnetic basis metal. They involve the measurement of the force of attraction of a permanent magnet to the ferromagnetic base, the thicker the coating between the usually rounded end of the magnet and the basis metal, the smaller the force of attraction.

A magnetic thickness meter can be designed according to either of two different principles. In the first design, a small magnet is attached to one end of a balance which is an equilibrium, the magnet being connected through a spiral spring with a micrometer screw or calibrated disc. To carry out the measurement, the spiral spring, which is adjustable, is stretched by turning the micrometer screw until the magnet is pulled away from the coated surface. The coating thickness can then be read off from the calibrated disc or obtained from a calibration curve. Instruments of this type include the Mikrotest manufactured by the Elektro-Physik, Cologne (Fig. 26), the Inspector Thickness Gauge manufactured by Elcometer Instruments Ltd. (Fig. 27) and the Magne-Gage made by the American Instrument Company (Fig. 28).

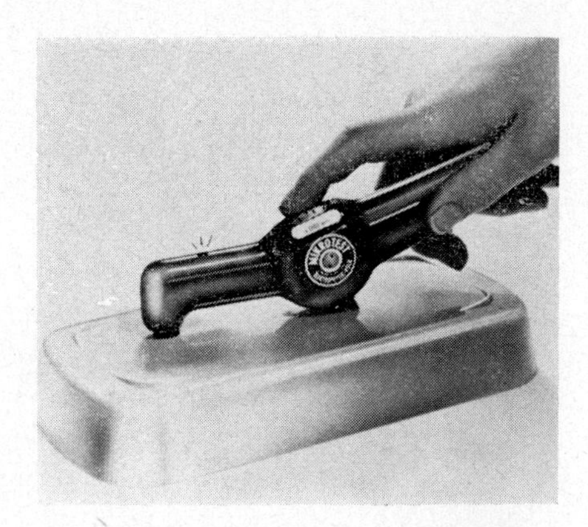

Fig. 26. Mikrotest magnetic thickness gauge.

For further details see Table 6, p. 104, No. 43.

Fig. 27. Inspector thickness gauge, Model III.

For further details see Table 6, p. 102, No. 28.

Fig. 28. Magne-Gage magnetic thickness tester.

For further details see Table 6, p. 102, No. 35.

An instrument for the checking of coatings on a ferrous substrate where a go-no-go indication is required is the Chemigage manufactured by Elcometer Instruments Ltd., Fig. 29. The instrument is placed on a calibrated foil of the required nominal test thickness or on a known good sample. An adjustable core is screwed up into the body of the instrument until the internal magnet remains in contact with the surface of the test piece. The core is then screwed down until the magnet just lifts from the surface. The instrument can then be applied to the coated surface to be tested. If the magnet remains down, then the coating thickness is below the original calibration foil or test sample. If it rises again, the coating thickness is either the same as the original or greater.

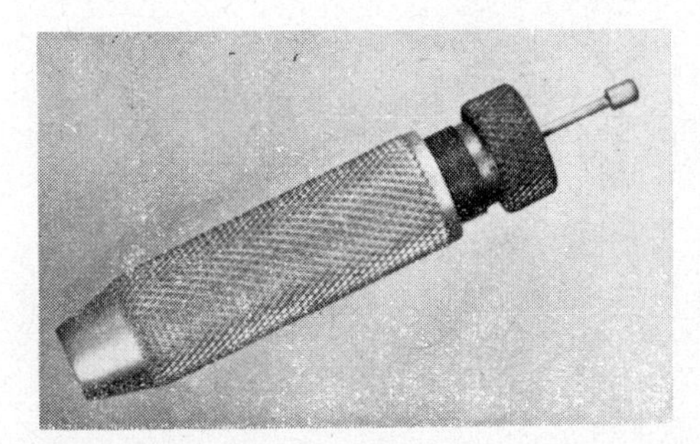

Fig. 29. Chemigage (Elcometer Model 107).

For further details see Table 6, p. 98, No. 9.

The other design comprises a rod-shaped permanent magnet with a hemispherical end. The magnet is provided with a scale and is suspended by a tension spring inside a pencil-shaped tube. The use of instruments of this type is extremely simple. The instrument is positioned vertically in contact with the test surface. The spring is then tensioned by drawing the outer tube vertically upwards until the resulting force is sufficient to overcome the force of attraction and pull the magnet away from the test surface. The scale reading is taken at the point at which the magnet is pulled away as in the case of the BSA-Tinsley pencil gauge manufactured by Evershed & Vignoles Ltd. (Fig. 30). If necessary, the test is repeated.

A similar device to the BSA-Tinsley gauge is the Elcometer Pull-Off Gauge illustrated in Fig. 31. The pull-off point of the magnet is indicated on the three scales calibrated in 0-600 microns, 0-25 thou and 0-100 arbitrary divisions.

The magnetic attraction thickness measuring instrument can obtain an accuracy of about ± 10%. Among the sources of error that must be taken into consideration as likely to have a considerable influence on the results is the shape of the specimen, i.e. whether the surface is concave or convex, the permeability and thickness of the basis metal if under 0·4 mm thick, the surface roughness and, in the case of soft materials, the effect of a purely mechanical sticking of the magnet to the surface.

Fig. 30. BSA-Tinsley Pencil Gauge magnetic thickness tester.

For further details see Table 6, p. 98, No. 7.

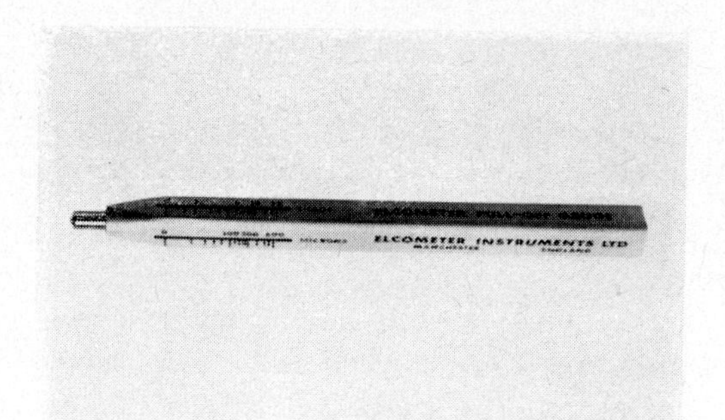

Fig. 31. Elcometer Pull-off Gauge, Model 157.

For further details see Table 6, p. 98, No. 15.

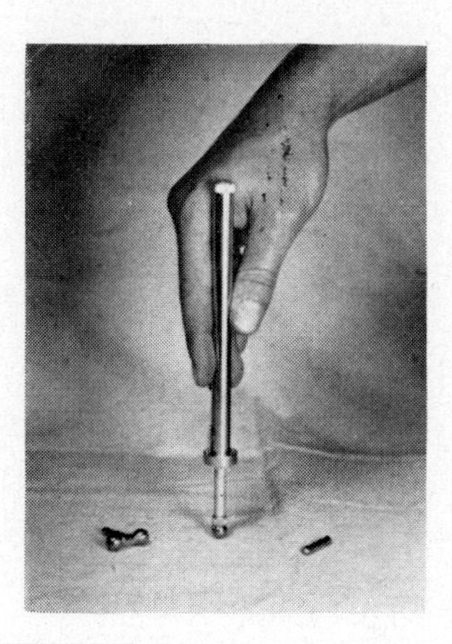

Fig. 32. MA S/P magnetic thickness gauge.

For further details see Table 6, p. 104, No. 37.

The range of application can be extended by altering the magnetic probe. The MA S/P[21] (Fig. 32) and the Magne-Gage instruments are accordingly supplied with various replaceable magnet inserts.

All the above instruments can be used for measuring the thickness of thin nickel coatings in spite of the fact that the latter are also ferromagnetic. This is done by making use of reference curves or comparative measurements. The Magne-Gage permits measurements on nickel coatings even up to a thickness of 50 μm.

Another magnetic gauge, the Lectromag (Fig. 33) by the Lea Manufacturing Co., makes use of the force of magnetic attraction in a manner different from that of the previously-described instruments and is only suitable for the measurement of non-magnetic coatings on steel. It consists of a glass tube carrying an engraved scale. Around this tube is arranged a vertically moveable coil connected to a source of a.c. The a.c. required is 110-115 V at 60 Hz. Inside the glass tube is arranged a vertically moveable iron core. The instrument is placed vertically on the test surface, the end of the iron core is then placed in contact with the surface and the coil is then slowly raised. At first the iron core, which is magnetised by the coil, will remain attracted to the surface but a point is reached at which vibration of the core caused by the a.c. potential indicates the imminent pulling away of the core. When this has happened, the point on the scale indicated by the marker on the top end of the core is read off, the core being held in equilibrium by the coil. The accuracy obtainable with this instrument is ± 10-15%.

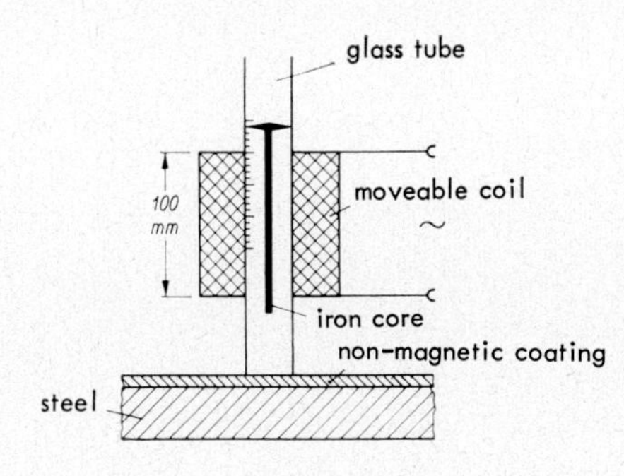

Fig. 33. Principle of the Lectromag.

For further details see Table 6, p. 102, No. 30. List of manufacturers, p. 119, No. 35.

21. Berthold, R. A handy coating thickness meter of high accuracy. *VDI-Z*, 1949, 91, No. 18, 476-478.

Magnetic induction methods

Magnetic induction methods can be used to measure the thickness of a non-ferromagnetic coating on a ferromagnetic substrate and also, under certain conditions, *vice versa.* The instruments used take various forms. The conductivity of the coating is usually of secondary importance. The influence of the permeability of the basis material, the shape of the test sample, and not least the design of the probe are, however, all of considerable importance as regards the accuracy of the measurements. An accuracy of $\pm$ 2 to 10%, depending on the design of the instrument, is attainable provided adjustment at two points is available. This means that a careful calibration of the instrument is required, first on the basis material without a coating or on a specimen corresponding exactly to the basis material, and secondly on a coating of known thickness lying within the range of values to be measured. An appropriate procedure must be adopted for ranges for which the starting point on the scale is other than zero.

The simplest forms of instrument belonging to this group are those which make use of the magnetic flux of a permanent magnet parallel to the test surface (Fig. 34). Examples include the Magnus Junior made by A. Bergner & Co. (Fig. 35) and Elcometer Thickness Gauge model 101 manufactured by Elcometer Instruments Ltd. (Fig. 36).

Essentially these instruments comprise a U-shaped permanent magnet the poles of which terminate in soft iron contact pieces. A subsidiary circuit is formed between the legs of the magnet by a small moveable magnet provided with a pointer. After the device has been placed on a magnetic substrate the magnetic circuit is closed through the contact pieces and the moveable magnet carrying the pointer takes up a certain position which corresponds to zero coating thickness. If now a non-ferromagnetic material is introduced between the contacts and the substrate, the magnetic flux passing through the contact pieces is reduced, and as a result the deflection of the moveable magnet in the subsidiary circuit is increased. The magnitude of the deflection provides a measure of the coating thickness and can be read off on a non-linear scale.

Other sub-divisions of the magnetic induction group of methods are the electrical induction, transformer and magnetic amplifier methods. They are used almost exclusively for the measurement of the thickness of non-magnetic coatings on ferromagnetic substrates.

Three basic arrangements have been developed. The electrical induction type[22] makes use of the changes in self induction and in the loss angle for

22. Deutsch, V. New developments in the field of magnetic coating thickness measurement. *Elektronik,* 1959, **8**, No. 6, 187-188.

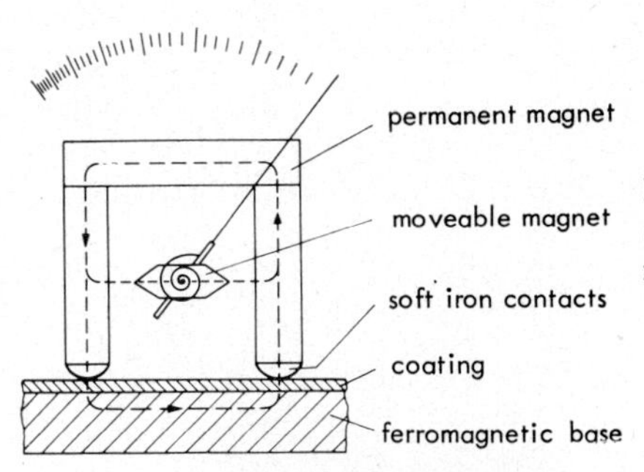

Fig. 34. Principle of magnetic induction instrument using a magnetic flux parallel to the surface.

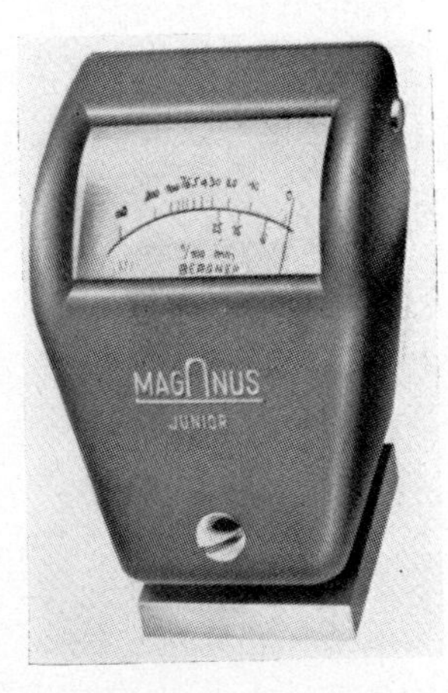

Fig. 35. Magnus Junior.

For further details see Table 6, p. 104, No. 36.

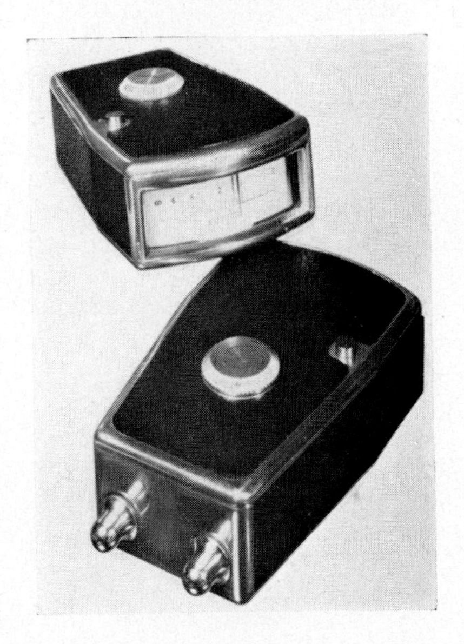

Fig. 36. Elcometer Thickness Gauge, Model 101.

For further details see Table 6, p. 100, No. 16.

measuring the coating thickness. Depending on the distance of the ferro-magnetic material from the probes, i.e. the coating thickness, the closure of the magnetic circuit, which forms part of a non-amplifying bridge circuit, is more-or-less complete, and the deflection of the pointer of the indicating instrument is proportional to the coating thickness. An example of the circuit of an instrument of this type and of the different probe arrangements is shown in Fig. 37. Probe A has two pole contact pieces, probe B has a point and a ring-shaped contact piece, while probe C has a single point of contact, the second ring-shaped pole being set back a certain distance above the surface. The use of the non-amplifying bridge circuit has several advantages such as insensitivity to voltage fluctuations and changes in frequency and to curve shape of the supply current. In addition, the calibration of the curve to the coating thickness in microns is practically linear.

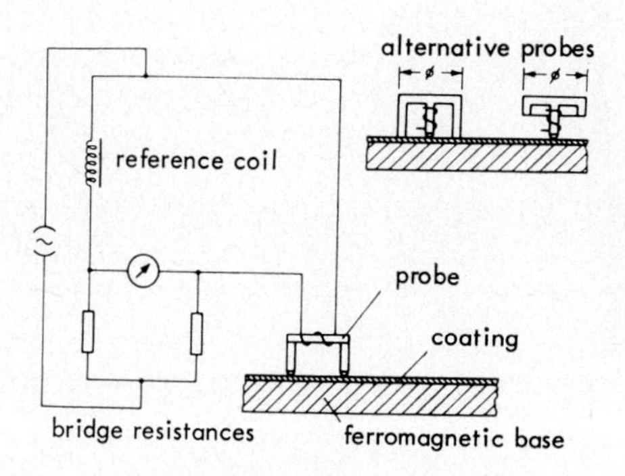

Fig 37. Circuit diagram of electrical induction method and examples of alternative probes.

The transformer method is based on the mode of operation of a magnetic amplifier. It is shown diagramatically in its simplest form in Fig. 38. The device comprises a small U-shaped iron core carrying a primary and a secondary winding. The primary winding is connected via a transformer to the a.c. mains. If the magnetic circuit is closed by a ferromagnetic substrate the potential across the secondary winding increases. This potential can then

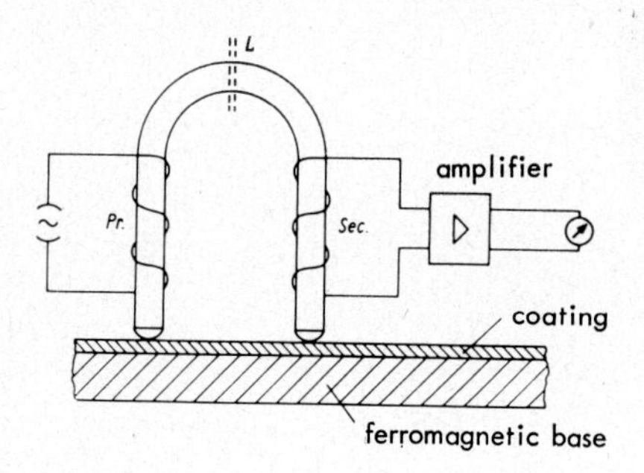

Fig. 38. Basic circuit of device based on the transformer principle. L represents an air gap.

be passed on through a normal a.c. amplifier to an indicating instrument. The magnitude of this potential varies depending on the thickness of the coating between the substract and the contact pieces.

Incorrect measurements may arise as a result of contact being made by only one of the poles of the probe. This may be avoided in the case of conducting coatings by modifying the core and applying a direct current. For this purpose the core is split into two halves insulated one from the other by the air gap L which, however, does not interrupt the magnetic flux. A direct current applied to both limbs of the core can be used to control the a.c. amplifier or to cause a signal lamp to light up when both poles of the probe are in contact with a conducting coating.

In the magnetic amplifier[23] (Fig. 39) the double core carries a measuring coil and an exciter coil supplied with d.c. which produces the permanent magnetic field required for the measurement. The measuring coil is connected to an a.c. supply and is influenced by the magnitude of the magnetic flux. The potential developed across the resistance is passed through an amplifier to an indicating instrument from which the thickness of the coating can be read off.

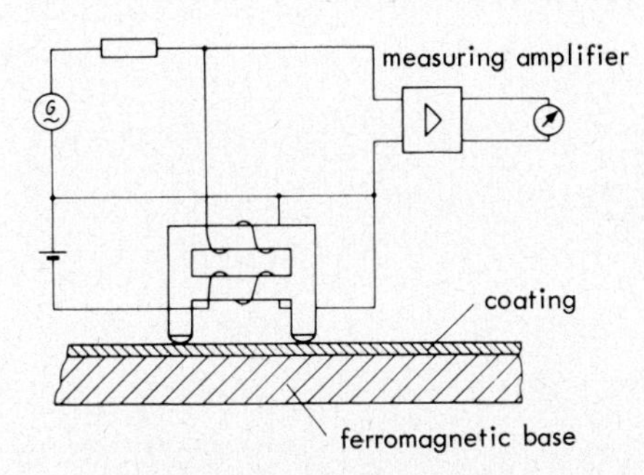

Fig. 39. Principle of the magnetic amplifier.

The principle just described is used in numerous instruments including the three manufactured by Karl Deutsch for different measurement ranges and methods of operation. They are the Leptoscope SMG type EP (p. 102) which has two measurement ranges, the Leptoscope Universal SMG 8 (Fig. 40) with four measurement ranges and the battery operated transistorised Leptoscope T 500 (Fig. 41). All three instruments are provided with a single pole probe, the Leptoscope SMG 8 using a two-pole probe for the largest measurement range of 1-8 mm. For the wider measurement ranges with scale deflections of up to several millimetres two-pole

23. Fischer, H. and Rupp, H. Non-destructive coating thickness measurement with the help of magnetic methods. *Z. Metallkde,* 1963, **54**, No. 6, 339-345.

probes are generally preferred for use with electrical induction instruments. The distance between the poles must not be less than a certain value. The two-pole probes give a more accurate signal and one which can be better evaluated. On the other hand, single pole probes (see Fig. 37, top right) enable measurements to be made on very small surfaces.

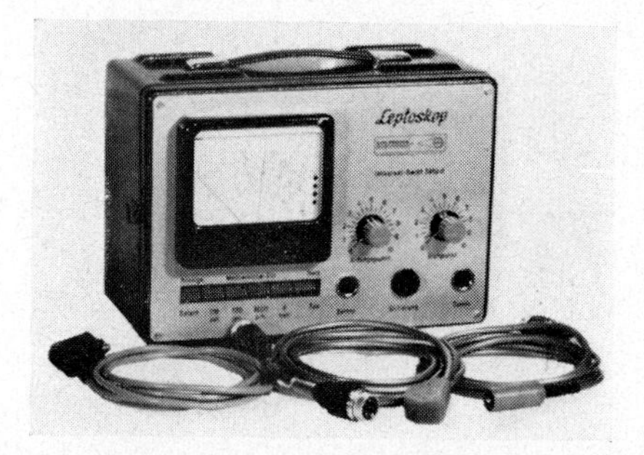

Fig. 40. Leptoscope Universal SMG 8.

For further details see Table 6, p. 102, No. 34.

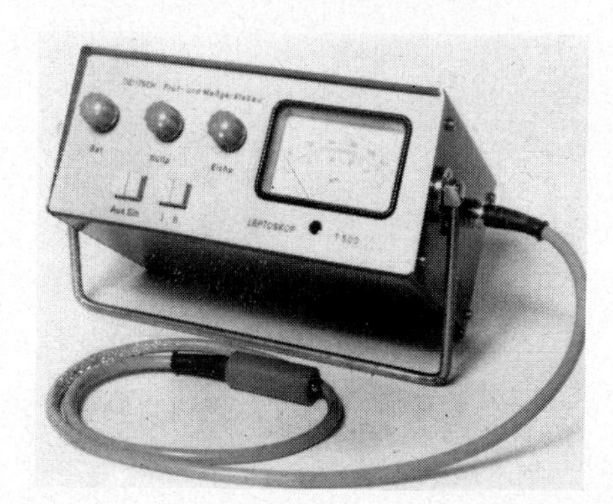

Fig. 41. Leptoscope T 500.

For further details see Table 6, p. 102, No. 33.

The Institut Dr. Förster offers a magnetic-induction thickness measuring apparatus known as the Monimeter type 2·094 (Fig. 42). It is transistorised and fed by eight 1.5 V. cells. The equipment is provided with a single pole probe which is largely independent of angle of application. The contact point can be exchanged by a few simple manipulations. The Monimeter is provided with four switched measuring ranges between 0 and 3000 μm. It permits the measurement of the thickness of non-ferromagnetic layers on ferromagnetic bases. The accuracy of measurement on all ranges is better than $\pm$ 3% of full-scale but is reduced by inhomogeneities and roughness of the base material surface, especially on the most sensitive range.

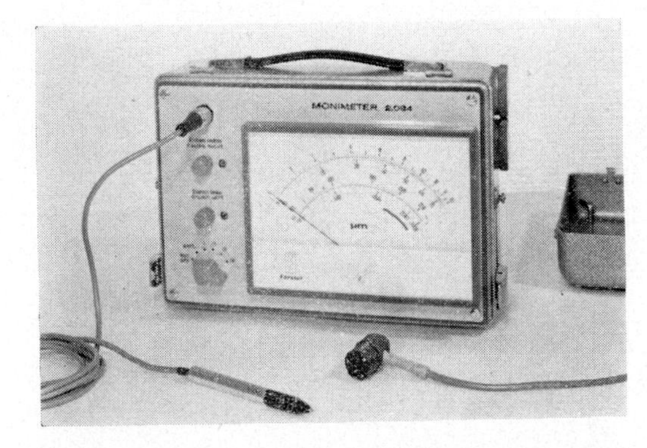

Fig. 42. Monimeter Type 2.094.

For further details see Table 6, p. 106, No. 48.

Elcometer Instruments Ltd. produce the Minitector Type G Model 158 with single point contact for the measurement of non-magnetic coatings on a ferrous substrate (Fig. 43). The instrument operates on the principle that when the tip of the probe is brought in close proximity with a ferrous material, the magnetic field of the tip is concentrated and intensified. Consequently, this increase in magnetic field is an indication of the distance the probe tip is from the ferrous substrate. Therefore, providing the coating is non-ferrous, the probe will detect the change in magnetic field due to the introduction of the coating. The signal from the probe is amplified and displayed on a meter in either imperial or metric units in four ranges from 0 to 0·2 inches (0 - 5 mm).

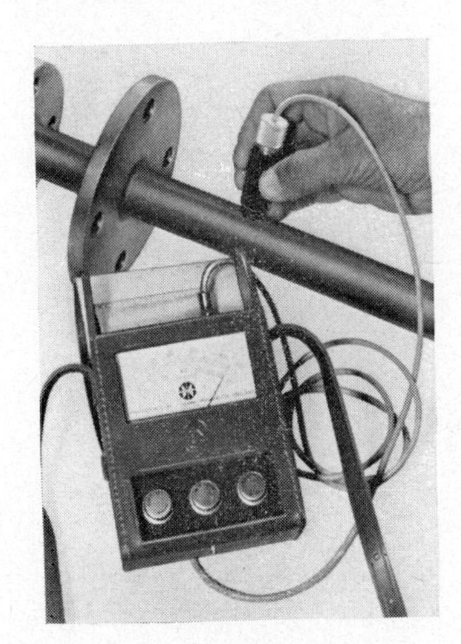

Fig. 43. Elcometer Minitector Type G Model 158.

For further details see Table 6, p. 104, No. 45.

The principle of the magnetic amplifier is used in three very similar instruments, the Permascope ES4[24] (Fig. 44) manufactured by Helmut Fischer G.m.b.H. and Co., the Elmicron FE (Fig. 45) manufactured by EFCO Ltd. and the Accuderm Model M (Fig. 46) manufactured by Unit Process Assemblies Incorporated. Each of these instruments uses a two pole probe for the measurement of non-magnetic coatings on ferrous bases. They each have three overlapping thickness ranges which may be selected from 0-0·1 inches (0-25 000 μm). For non-portable use a mains supply of 110/220/240 V at 50-60 Hz is required and for portable use a mains-independent supply by rechargeable nickel cadmium batteries is provided. Special probes are available for the measurement of coatings inside tubes and other special applications. The accuracy of indication is 3%.

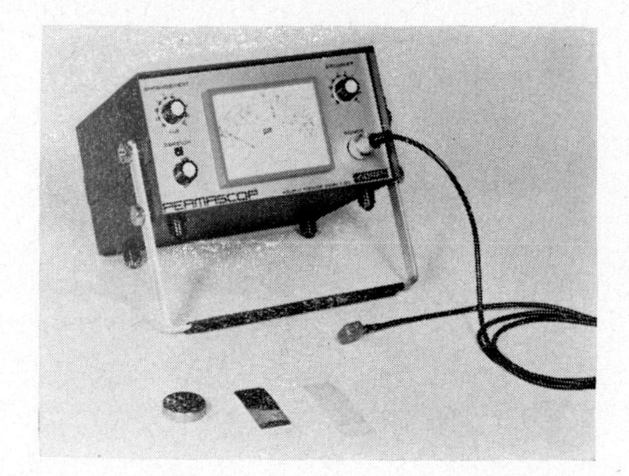

Fig. 44 Permascope Type ES4.

For further details see Table 6, p. 106, No. 55.

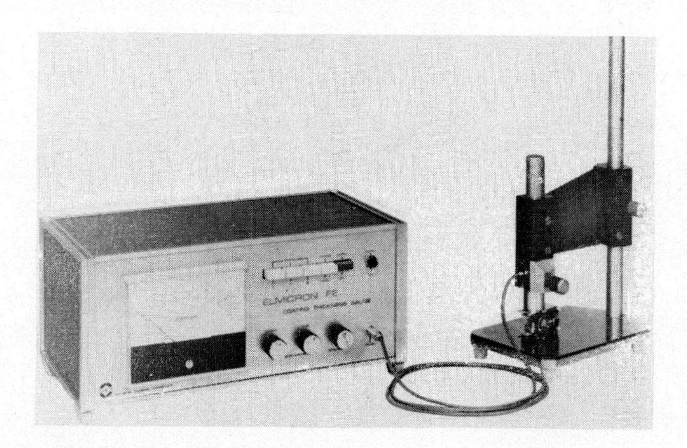

Fig. 45. Elmicron FE.

For further details see Table 6, p. 100, No. 45.

---

24. Gühring, H. The application of non-destructive coating thickness measuring methods. *Galvanotechnik,* 1967, **58**, No. 2, 109-111.

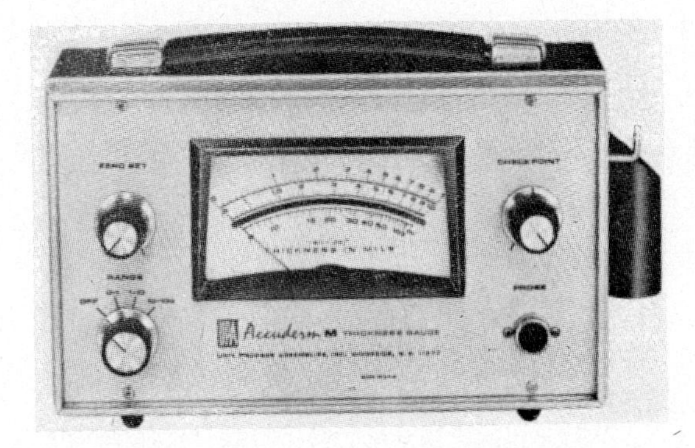

Fig. 46. Accuderm M.

For further details see Table 6, p. 96, No. 1.

A variation of the Accuderm Model M is the Accuderm D with three selectable ranges and direct digital readout in thickness units.

The electromagnetic layer thickness measuring instruments of the firm Elektro-Physik carry the name Elektrotest. The standard Elektrotest (Fig. 47) has a single pole inductive probe. It has only a single measuring range which can, however, be supplied to order for any scale between 0 and 10 000 μm. The equipment is intended for connection to 220V a.c. mains. It permits measurements of non-magnetic deposits on ferrous bases.

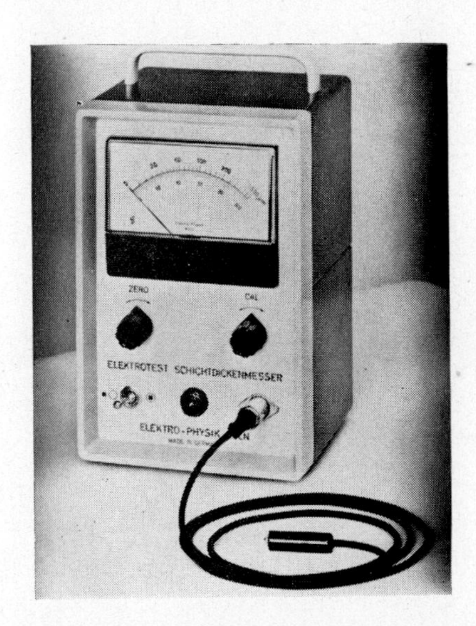

Fig. 47. Elektrotest Standard model.

For further details see Table 6, p. 100, No. 19.

For thickness measurements of insulating layers on non-ferrous metal bases, the Elektrotest Type U is available (Fig. 48). This apparatus is also intended for connection to 220V a.c. mains.

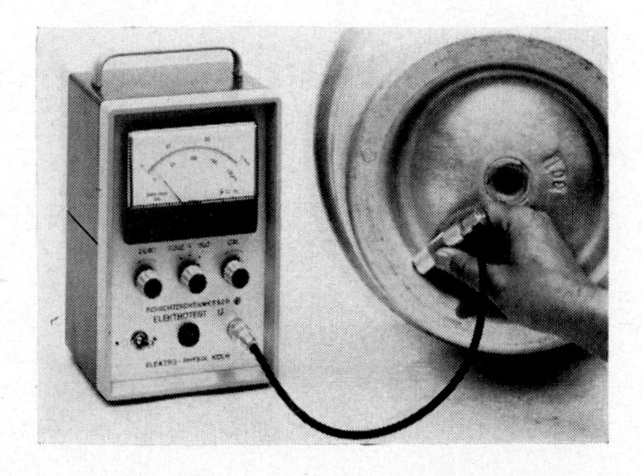

Fig. 48. Elektrotest Model U.

For further details see Table 6, p. 100, No. 20.

The Elektrotest Model FP (Fig. 49) is a mains-independent apparatus provided with three readily interchangeable batteries each of 9 volts. It has two switched ranges for non-ferromagnetic coatings which are selectable between 0 and 2 500 μm.

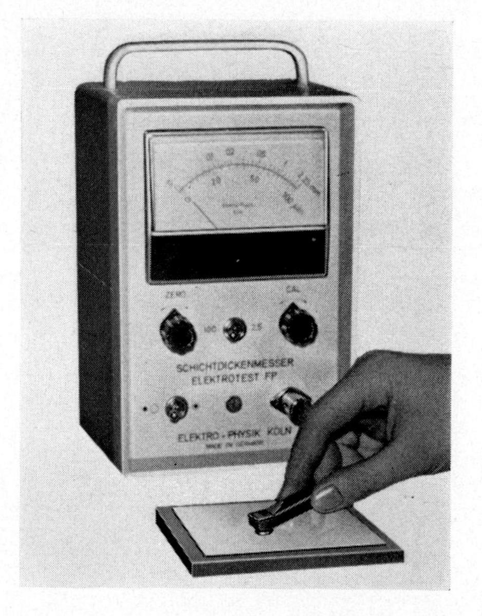

Fig. 49. Elektrotest Model FP

For further details see Table 6, p. 100, No. 18.

The measuring accuracy of all Elektrotest equipments is given as 5% of the measured value. The principle of operation of these instruments can be seen from Fig. 50. Basically it is similar to that of instruments operating on the basis of a magnetic flux parallel to the surface, already shown in Fig. 34. The difference is in the arrangement of the magnetic shunt. A soft iron component lies above the system, parallel to the permanent magnet. Between the soft iron member and the permanent magnet there is a semi-conductor whose electrical resistance depends upon the strength of the magnetic field. When the soft iron pole pieces approach a ferromagnetic base, the external magnetic flux increases so that the semi-conductor decreases in reverse ratio. An instrument calibrated in μm shows the resistance of the semi-conductor, which is a function of the distance between the soft iron pole pieces and the ferromagnetic base, and thus the thickness of coating being measured.

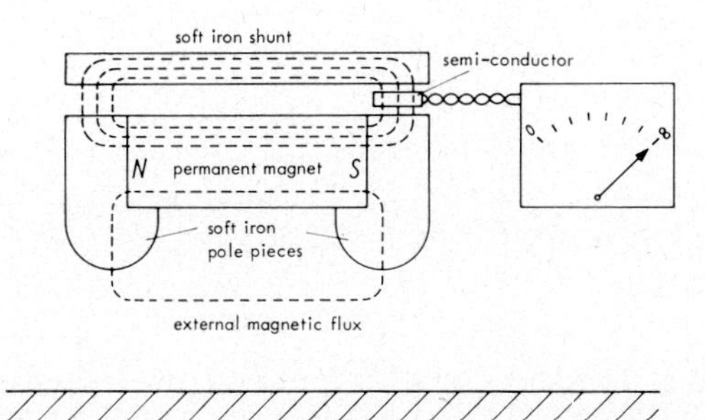

Fig. 50. Functional diagram of the Elektrotest FP.

Also to be mentioned is the Reluctance-type Thickness Gage (p. 108) of General Electric Company. This is an instrument for measuring the thickness of non-ferromagnetic layers on iron. The measuring range is selectable between 0 and 750 μm. Power is by a.c. at 50 or 60 Hz. The apparatus operates with a tranformer in the saturation region which steps down the mains voltage to the voltage needed for the bridge circuit. For measurement, the coil of the test probe, which is in one arm of the bridge, is applied to a check sample with a known coating thickness. By means of a potentiometer a pointer instrument is set to read exactly the known thickness of this coating. Coating thicknesses whose values lie within $\pm$ 50% of this setting value can then be determined with an accuracy of 5 to 10%.

Recent additions are the SM 1, SM 2 and SM 3 coating thickness meters and the Diameter coating thickness meter of Dipl. Ing. Heinrich List. In these instruments, the flux change of a permanent magnetic field brought about by the interposition of the coating to be measured is conveyed by a field plate (a magnetically-sensitive semi-conductor) to an indicating instrument. The probe is joined to the instrument by a flexible cable and the positioning of the probe on the test surface gives an immediate pointer reading with an accuracy of $\pm$ 5%. A calibration plate is provided and adjustments are included to compensate for variations in temperature and in battery voltage. The different models have different probes to cover different ranges of coating thickness.

Eddy current methods

Perhaps one of the most interesting procedures for the non-destructive measurement of coating thickness employs the eddy current principle. Equipments which have been developed to apply the process have an extraordinarily wide range of application and can be applied to practically all likely metallic and insulating coatings on any base. There is a limitation in that the conductivity or, in the case of ferromagnetic materials, the permeability of the layer must differ sufficiently from that of the base. The greater this difference, the better the accuracy of measurement.

Fig. 51, taken from the description of the Dermitron apparatus which will be described in more detail later, shows the ability to measure material combinations as a function of electrical conductivity. For the coating and base metals, only the so-called linear reading has to be determined. The values appropriate to the two metals are read off from the curve, irrespective of which is the coating and which the base. The algebraic difference of their values on the abscissa is read off, and if this is greater than 100 a satisfactory thickness measurement will be possible. Thus, a layer of silver on nickel silver is

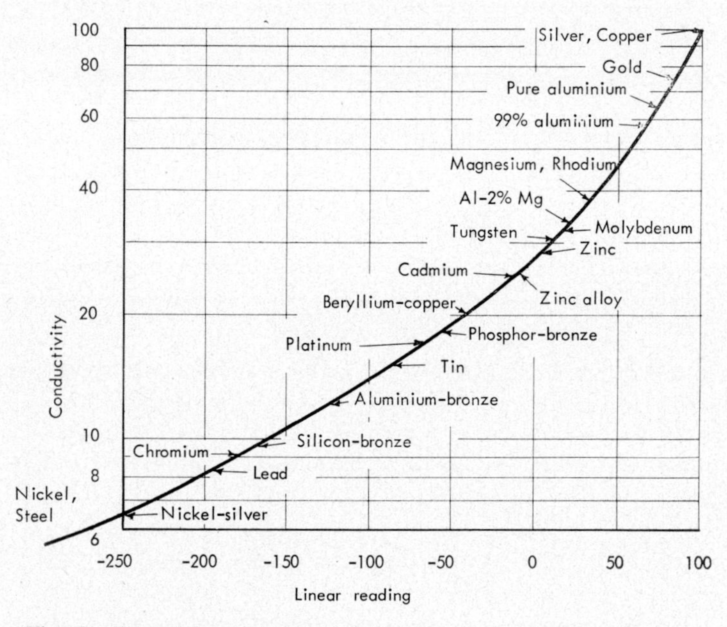

Fig. 51. Material combinations which can be measured by the eddy current method (Dermitron).

measureable but one of silver on copper is not. It should be mentioned that the possible measurement of nickel coatings on steel cannot be deduced from the diagram.

If a coil carrying high frequency alternating current is applied to the coating to be tested, then an eddy current is generated in the adjacent metal; this has a reactive effect on the primary current in the coil from the effect of induction, and so influences the impedance of the measuring coil which forms part of a bridge circuit (Fig. 52). The out-of-balance of the bridge which results is dependent on the layer thickness which separates the coil from the base metal, and this is indicated by the deflection of a pointer in the measuring apparatus. The depth of penetration of the high frequency currents, which can be at up to 14 MHz, is limited by the skin effect.

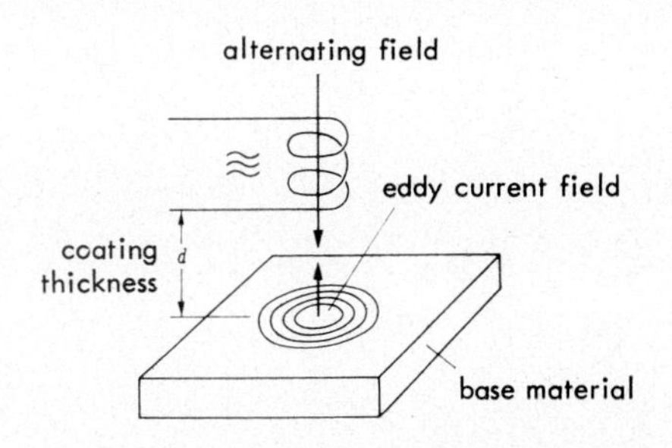

Fig. 52. Principle of the eddy current method.

A universal apparatus for all measurement problems which can be solved by application of the eddy current effect would be elaborate. For this reason some equipment manufacturers only take into account problems which cannot be solved by the application of other procedures.

A very versatile thickness measuring apparatus operating on the eddy current principle is the Dermitron already mentioned[25, 26] (see Fig. 53) of Unit Process Assemblies Inc. The equipment has four switched frequency measurement ranges between 0·1 and 6 MHz; this offers a variety of measurement possibilities. The thickness of metallic and non-metallic coatings on nearly any base metal can be determined (see Fig. 51). The measurement ranges vary, according to the material combination, between 0 and 2540 μm. The coating thickness can be read on the scale directly in μm or in thou. The scales are interchangeable.

25. Brodell, F. P. and Brenner, A. Further studies of an electronic thickness gage. *Plating*, 1957, **44**, 591.
26. Riccio, V. A non-destructive method for the measurement of the individual coatings thickness in the copper and nickel electrolytic deposits on ferrous base. *Plating*, 1962, **49**, 862-869.

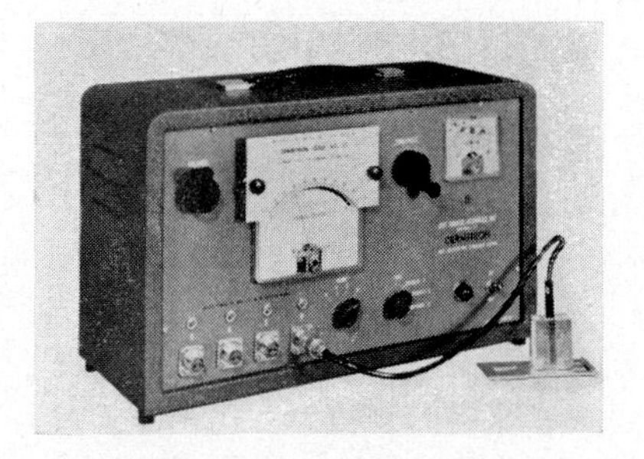

Fig. 53. Dermitron.

For further details see Table 6, p. 98, No. 12.

Fischer Instrumentation (GB) Ltd. currently supply several types of transistorized instruments operating on eddy current and electro-magnetic induction principles. The Permascope type ES is used for measurement of all non-magnetic coatings, whether metallic or not, on ferromagnetic substrates. A specially designed probe concentrates the induced field and enables extremely small items to be measured. Standard instruments are available for measurements in the range 0·5 μm to 10 mm.

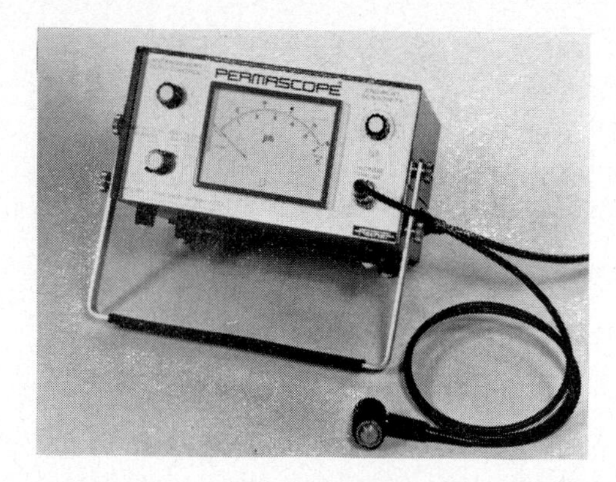

Fig. 54. Permascope type EC 3.

For further details see Table 6 p. 106, No. 54.

The Permascope EC3 eddy current instrument was primarily developed for non-conductive coating measurement on non-ferrous substrates. Coatings on flat or curved surfaces may be measured in the range 0·5 μm to 150μm, but instruments are available for coatings up to 70 mm. The new Isoscope is the pocket sized addition to the Fischer eddy current range and represents the first instrument of its type capable of measuring non-conductive coatings on both ferrous and non-ferrous substrates. This instrument is capable of measurement up to 350 μm.

The Nickelscope utilises a modified electro-magnetic induction technique in order to eliminate the inherent variables encountered when measuring nickel coatings. Calibrated using standards produced from the nickel process being tested, the Nickelscope measures nickel as foil or deposited on either ferrous or non-ferrous substrates, whether or not the nickel is coated in turn by say gold or chromium.

All the above mentioned instruments are available for mains only operation (except the Isoscope) or for portable use on batteries Each has numerous accessories available for undertaking diverse applications, including continuous measurement.

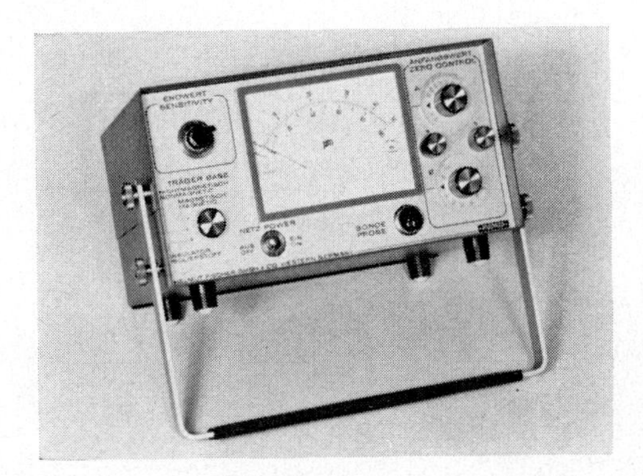

Fig. 55. Nickelscope Permascope type EN 1.

For further details see Table 6, p. 106, No. 50.

The Isometer Type 2·082 (Fig. 56) of the Institut Dr. Förster is a transistorised battery powered equipment. It measures the thickness of non-conductive coatings on non-ferromagnetic metals. With two switched measuring ranges of 0 to 100 and 0 to 300 μm practically all usual insulating coatings, such as lacquer, paint, and anodised layers, are covered. The associated normal tripod measuring head requires a level measuring surface of at least 20 $\times$ 20 mm^2.

Fig. 56. Isometer type 2.082.

For further details see Table 6, p. 102, No. 29.

A pocket model eddy current instrument with a weight of only 450 g. is the newly developed Leptometer Type NET 200 (Fig. 57) of Karl Deutsch. In accordance with the principle of measurement, the thickness of all non-conductive coatings on metals can be determined with it. With a single pole miniature measuring probe, accuracies of 10% of the measured value are attained. The normal range is 0 to 200 μm. The instrument is powered by a 9 V battery or a rechargeable accumulator. Four different measuring probes are available for different base metals.

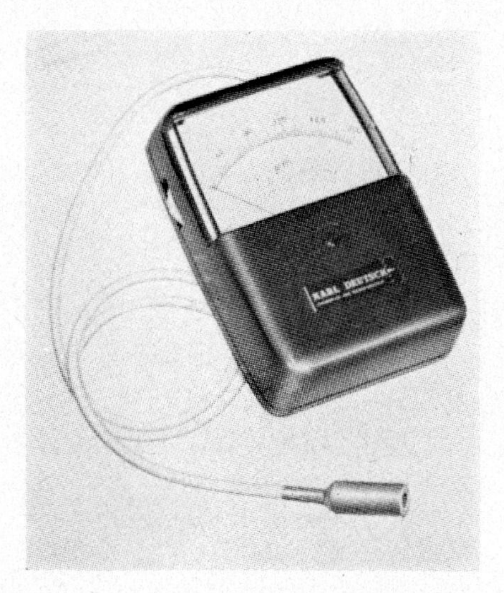

Fig. 57. Leptometer type NET 200.

For further details see Table 6, p. 102, No. 31.

A manufacturer of three further coating thickness meters operating on the eddy current principle is Elcometer Instruments Ltd. The Minitor Thickness Gauge* is also a transistorised pocket instrument, of which four models are available. The first is a standard model with tripod probe, the second for round samples with a prismatic probe, the third for soft coatings using a flat probe, and the fourth for deep-lying measurement surfaces with a special extension. Six scales with ranges of between 0 and 5000 μm are offered as standard. The smallest measuring range is 0 to 100 μm.

---

* Since this was written we have been informed that the Minitor Thickness Gauge Model 102 has temporarily been discontinued. It is being redesigned and should be available again later in 1971.

As opposed to the Minitor gauge, which is only suitable for non-conductive layers on non-ferrous bases, the Elcotector Mk. III (Fig. 59) is a versatile instrument. Switched ranges and a variety of interchangeable probes permit the determination of thickness of conducting or non-conducting deposits on ferrous and non-ferrous metals with the limitation for the eddy current method already mentioned that base and deposited metal must show a sufficient difference in conductivity. The Elcotector, like the Minitor, can be powered by a battery. It is also available as a mains powered instrument for 110 to 240 V a.c. for 50 and 60 Hz.

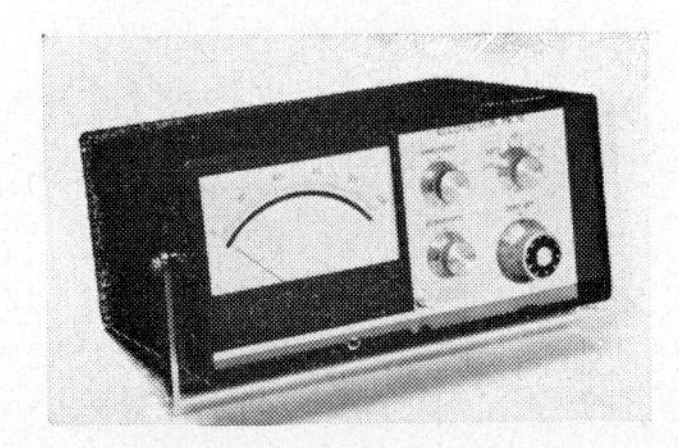

Fig. 59. Elcotector Mk. III.

For further details see Table 6, p. 100, No. 17.

The Eddy Gauge Model 133 (Fig. 60) is similar in appearance and operation to the Elcotector Mk. III. It is, however, completely portable with a nickel-cadmium rechargeable battery; for internal use it may be connected to an a.c. supply of 115-230 V 50-60 Hz. The instrument is intended for the measurement of non-conducting coatings on non-ferrous bases in three switched ranges from 0-250 μm or 0-0·010 inch with a general accuracy of $\pm$ 10%.

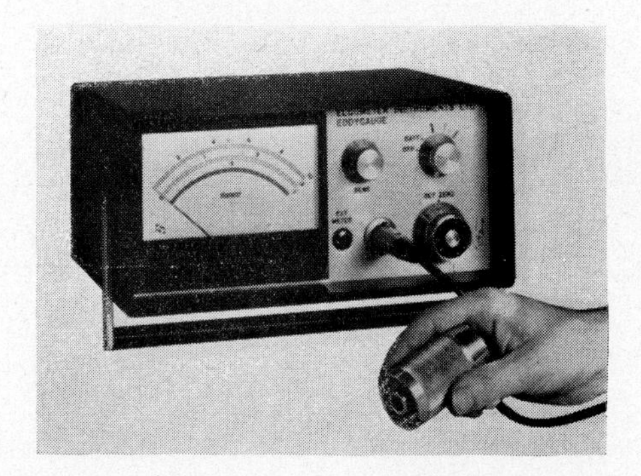

Fig. 60. Eddy Gauge Model 133.

For further details see Table 6, p. 98, No. 14.

The American Instrument Co. supplies the Filmeter* NRL Model 1 (p. 100) which has remained unchanged for some years. The Filmeter, a

* Stop Press: now being discontinued.

battery powered instrument, also serves for the determination of the thickness of insulating coatings on non-ferrous metals. The method of operation differs from that of the other eddy current instruments. Two frequencies, one constant and the other variable, are compared with each other using a pair of headphones as indicator. With the probe applied to the uncoated base material, the variable frequency is set to the zero position, i.e. with the variable frequency equal to the constant one. On applying the probe to the coated surface, the beat frequency changes in accordance with the coating thickness. By turning a control carrying a 100 unit calibration, the frequencies are again brought to equality. From the setting of the scale the thickness is then read off on a calibration curve. The accuracy of measurement is around 5%.

The Ultrasonoscope Film Thickness Meter* (Fig. 61) is available in two models. Type ASF/11 measures anodic oxide coatings on all aluminium alloys in the range 0 to 50 μm. Type PCF/1 is for paint coatings on non-ferrous metals, with a measuring range of 0 to 400 μm. The special feature of these instruments is the extremely low error of measurement. For type ASF/1 an absolute error of only ± 0·25 μm is quoted, and for the second type an error of ± 2·5 μm. These instruments can be supplied to operate from a.c. current of 90 to 240V at 50 or 60 Hz.

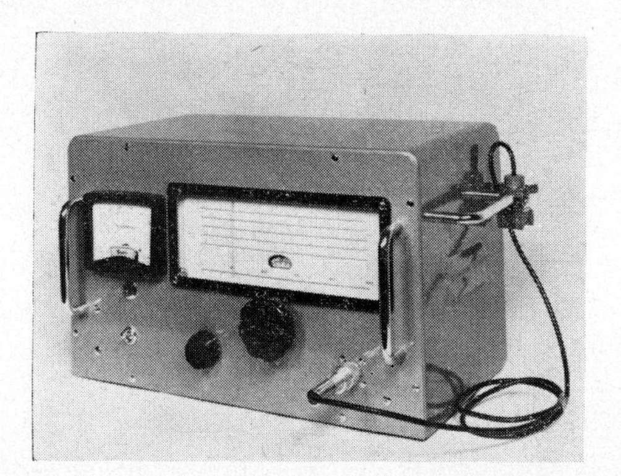

Fig. 61. Ultrasonoscope.

For further details see Table 6, p. 108, No. 61.

The Elmicron NF (Fig. 62) manufactured by EFCO Ltd. is a battery/mains portable instrument using rechargeable nickel-cadmium batteries. It is intended for measurement of all non-conductive films on non-ferrous bases in three overlapping thickness ranges selected by push buttons, 0-30 μm, 25-100 μm, 75-300 μm, or in imperial thickness units. General accuracy is 5%. Special measuring stands are available when very thin coatings are to be measured. When connected to the mains a.c. supply the internal battery is charged, the state of charge being shown on a separate meter.

* Stop Press: now replaced

Fig. 62. Elmicron NF

For further details see Table 6, p. 100, No. 22.

Capacitative methods

For the measurement of coatings with a high specific resistivity, such as insulators, it is possible to use the relatively-unknown capacitative thickness measurement method (see Fig. 63).

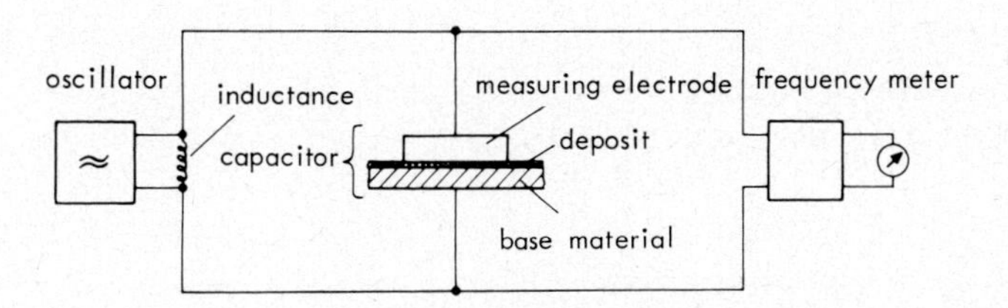

Fig. 63. Principle of the capacitative method.

An electrode, usually plane, of definite area forms a condenser with a conductive base material, the coating forming the dielectric. The condenser together with an inductance forms a turned circuit which is used to determine the frequency of an oscillator. A change in the dielectric, corresponding to a different thickness, results in a frequency change which can be applied to the determination of the thickness of the deposit.

Capacitative methods can be applied to hygroscopic materials only if a constant atmosphere can be provided, since a change in moisture content of the material of only 1% can introduce a measuring error of up to 40%. Under optimum conditions, which must be taken to include a uniform coating, thicknesses down to 0.01 μm can be measured with an accuracy of ± 5%

Thermoelectric method

This method (Fig. 64) makes use of the thermoelectric properties of the combination of coating metal and base metal. A heating probe at a constant temperature of 130°C is pressed on to the coating to be measured. At the junction of the coating and the base metal immediately under the probe a thermoelectric e.m.f. is generated which is a function of the layer thickness and the nature of the two metals.

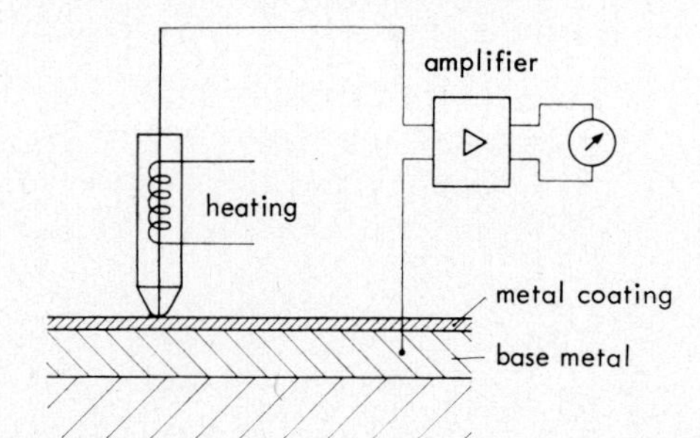

Fig. 64. The thermoelectric principle.

Any combination of metals can be measured provided that they develop a sufficiently high thermoelectric e.m.f., of the order of 100 μV, to permit measurement after passage through a d.c. amplifier.

At the present time there does not appear to be an instrument on the market operating on this principle. The B.N.F. Thermo-Electric Thickness Meter was manufactured and marketed for a time, first by Nash and Thompson Ltd. and then by their successors, Thorn Electronics (now Thorn Bendix Ltd.). It had the advantage of being a non-destructive instrument capable of measuring the thickness of nickel coatings on steel, zinc-diecastings and brass or copper, but it was found that its response varied considerably because different bright nickel coatings gave somewhat different thermoelectric effects, and that these were also affected by quite small concentrations (e.g. around 1% or less) of cobalt in nickel. Consequently it could only give reasonably useful results when the instrument was calibrated to suit each particular bright nickel plating bath being used. Although it has been employed for routine control in one or two large plating shops of manufacturing companies its use never became popular. It was, however, also a useful instrument for ‘metal sorting’ and rather more instruments were sold for that purpose than for thickness testing.

Optical methods

Optical methods for thickness measurement of opaque layers have the general disadvantage that an imperfection or step must be present or must be made. However, for the non-destructive measurement of thickness of transparent layers such as anodic oxide films, including coloured ones, several optical methods are suitable.

In the viewing method, the microscope is focussed first on the coating surface and then on the surface of the base. The thickness of the coating can then be deduced from the movement scale of the objective, taking the refractive index into account. Sharp focussing is, however, only possible if the surfaces show slight irregularities. Sharp focussing on the upper surface of the coating can be assisted by making a mark with a soft lead pencil and focussing on to the graphite particles.

Light section method

A much more elegant and accurate test is the light section method of Schmaltz[27], the principle of which is shown in Fig. 65. Rays from a monochromatic light source are imaged from a slit by the objective 1 and directed onto the surface of the coating at an angle of 45°. Part of the beam is reflected at the surface of the coating while part penetrates the coating and is reflected at the surface of the base material. Objective 2 makes visible two lines in the eyepiece, corresponding to these two different light paths. Since, on entering the layer, the rays are deviated in accordance with the refractive index of the material, this refractive index must be known for the evaluation

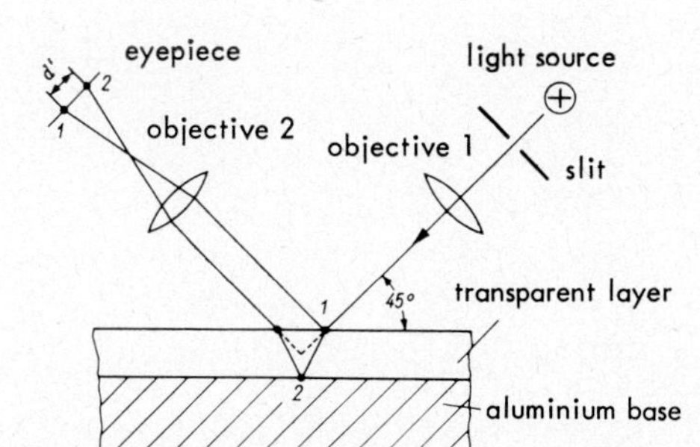

Fig. 65. Light section method.

27. Illig, W. The measurement of anodic coatings on aluminium. *Metalloberfl.*, 1959, 13, No. 2, 33-35.

of the measurements. The ratio of the true coating thickness d to the observed thickness d′ is given by the relation

$$d = d' . \sqrt{2 \cdot n^2 - 1}$$

Taking, as an approximation d = 2d′ (the refractive index of an anodised layer being about 1.59, and of compressed layers about 1·62), the error is only 3%.

Fig. 66 shows, in the upper part, the photographs of light reflections from a milky, nearly opaque anodised coating, and below those of a very clear transparent lacquer layer.

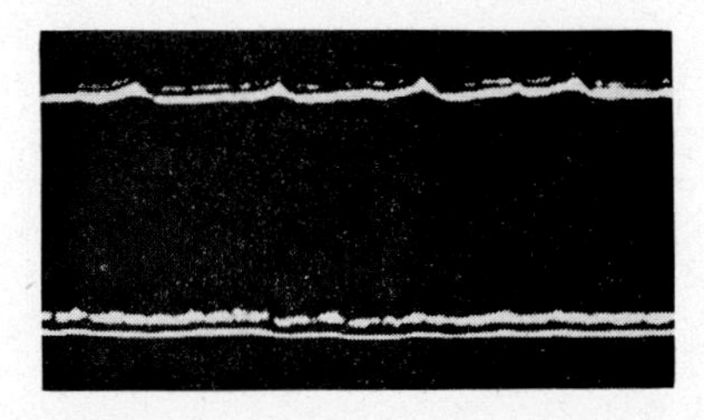

Fig. 66. Light section photographs of an anodised layer (upper) and a transparent lacquer coating (lower).

Another advantage of the light section method is that measurements can be undertaken with previous calibration provided the microscope magnification is known, which is of course usually the case. It is possible to make measurements on surfaces which are poorly defined and, besides the actual thickness measurement, an estimate can be made of the roughnesses of the surfaces of both the deposit and the base material. An instrument working on this principle is supplied by Carl Zeiss of Oberkochen (Fig. 67). A photographic attachment is available for recording purposes.

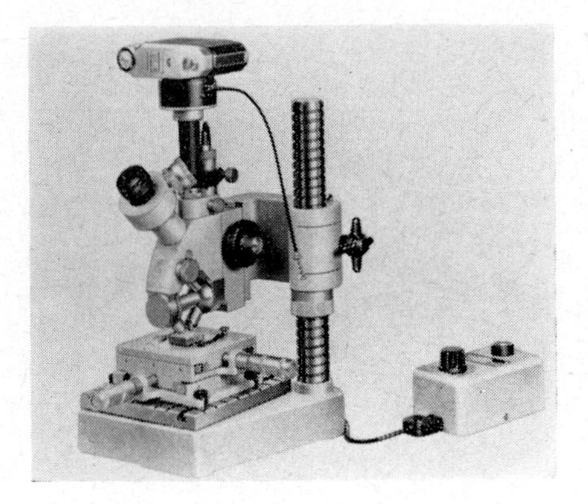

Fig. 67. Light section microscope.

For further details see Table 6, p. 104, No. 42.

Interference methods

Although the interference methods, like the light section method, are really processes for studying the structure or the form of surfaces, they can also be applied to the non-destructive measurement of thicknesses of transparent layers. Fig. 68 shows the schematic set-up of a two-beam interference microscope on the Michelson principle.

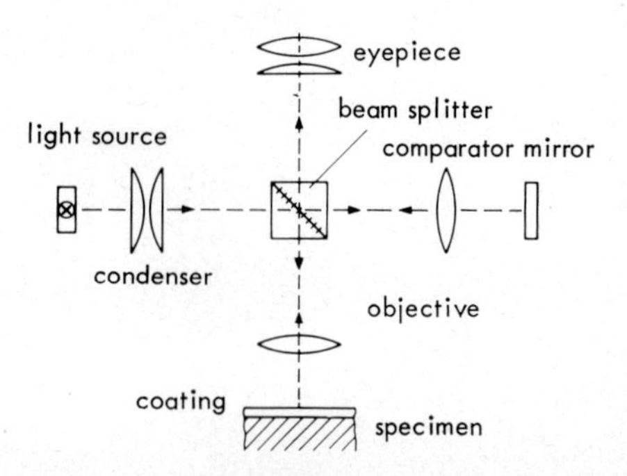

Fig. 68. Schematic set-up of a two-beam interference microscope.

Light from a source passes through an objective and falls on a half-silvered prism which splits the beam into two coherent beams. These two beams pass through similar objectives, one to a comparison mirror and the other to the specimen, from which they are each reflected back to the prism, where they form interference bands which are viewed in the eyepiece. Since the test specimen gives two reflections, one at the surface of the coating and one at the base, three sets of interference fringes are set up. The coating thickness can be deduced by counting the fringes and taking account of the known wavelength of the light and the refractive index of the coating. Interference microscopes of the Michelson type are supplied by Carl Zeiss (Fig. 69), Leitz/Wetzlar (Fig. 70), W. Watson, Microscopes Nachet and Hacker Instruments Inc.

Fig. 69. Zeiss double beam interference microscope.

For further details see Table 6, p. 104, No. 40.

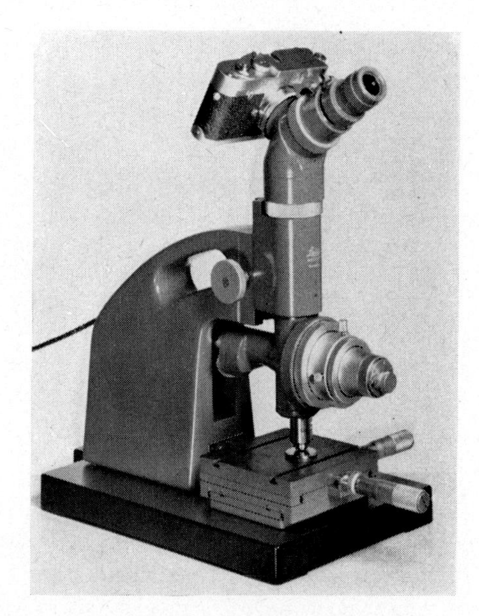

Fig. 70. Leitz double beam interference microscope.

For further details see Table 6, p. 104, No. 39.

Since interference microscopes are usually arranged for a choice of white or monochromatic light, another method, that of Elssner[28], can be used. In this method, the interference system is first set up in white light and the distance between two corresponding interference fringes is determined. After switching over to monochromatic light, the number of interference fringes between the same two fringes positions is counted, with the same setting. The thickness of the layer can then be deduced knowing the refractive index of the coating material (between 1·59 and 1·62 for an anodised layer) and the wavelength of the monochromatic light. The procedure is relatively accurate and takes only a few minutes. The smallest measurable thickness is below 1 μm.

Steps of about 50 Å (5×10^{-3} μm) upwards, which are formed at the edge of the coating by the application of extremely thin layers can be measured by the multiple beam interferometer method of Tolansky[29]. The measuring accuracy attainable on a flat base is around $\pm$ 0·001 μm independently of layer thickness.

For the generation of multiple beam interference[30], the so-called Fizeau

28. Elssner, G. The non-destructive determination of the thickness of anodic layers with the interference microscope. *Aluminium,* 1959, **35**, No. 4, 202-204.
29. Tolansky, S. Multiple-beam interferometry. Clarendon Press, 1948.
30. Dühmke, M. Coating thickness measurement with the help of optical interference. *Fachberichte für Oberflächentechnik,* 1966, **4**, No. 6, 205-206.

arrangement must be used which differs from the Michelson double beam arrangement in that the comparison mirror shown on the right of the Michelson diagram is moved down to the position of the specimen. In addition, the comparison mirror is silvered to give about 80% reflection, so that rays are reflected several times between the reflector and the specimen. Compared with the double beam system, the multiple beam arrangement gives an interference pattern having particularly sharp and narrow minima. Measurements on this principle can be made with a combination of a normal interference microscope and simple attachments which most firms supply. The interference microscope must be provided with a monochromatic light source (usually a sodium one) and should have only low magnification (about × 20). A range of interference and differential interferometry microscopes is available from Hacker Instruments Inc.

A method for non-destructive thickness measurement by the use of semiconductor techniques has been studied by Bogenschütz, Bergmann, and Jentzsch[31]. It depends on transmission and reflection measurements in the infra-red. Oxide-coated test speciments of silicon, germanium and tantalum were used in the investigation.

Photoelectric method

With the aid of a photocell the increasing coloration of anodised, vacuum deposited and other coloured coatings can be used as a measure of thickness. As seen in Fig. 71 an intense light beam is directed on the layer to be measured. The partially reflected light falls on a photocell whose output current is taken through an amplifier of suitable gain to an indicating instrument. The

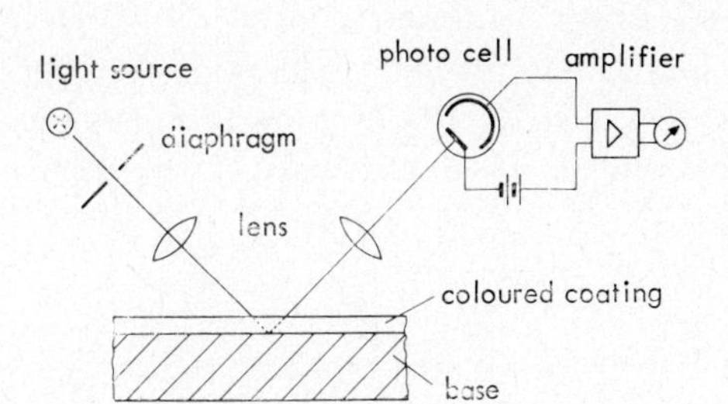

Fig. 71. Principle of the photoelectric method.

31. Bogenschütz, A. F., Bergmann, F. and Jentzsch, J. Non-destructive coating thickness measurement by an interference method. *Telefunken-Röhre,* 1963, No. 42, 159-176.

calibration of the instrument is carried out using standard test specimens or by drawing up a calibration diagram. The accuracy of this photoelectric method depends largely on the uniformity of the coating.

The optical film monitor of Edwards Instruments is primarily intended for the monitoring of thin film coatings during vacuum deposition but it is equally suitable as a static bench-mounted system for inspection, checking and grading of finished work. It can be used in reflection or transmission modes. The basis of operation is the sensing of the change in intensity of a modulated light beam after reflection by, or transmission through, a coated surface, wavelength separation being effected by the introduction of absorption or interference filters. The equipment includes a photometer head complete with a photocell for use in either the reflection or the transmission mode, a pre-amplifier and a main amplifier. With the installation of a second photocell and pre-amplifier, reflectance/transmittance comparisons can be made in rapid sequence.

The normal range of the instrument is between 10,000 (red) and 4,000 (blue) Å, but the addition of an infra-red accessory extends the longer wavelengths to 25,000 Å (near infra-red region). Beyond the limits of the optical system, monitoring can be extended to both the ultra-violet and deep infra-red regions by conversion of the mass/thickness relationship as determined by the Edwards film thickness monitor.

Modulation of the light beam and tuning of the amplifier to the modulation frequency ensures that the equipment responds only to the modulated light and is insensitive to light reaching the photocell from any other source. The range of sensitivity covers minimum light levels as obtained in anti-reflection coatings on glass surfaces to maximum levels as obtained in transmittance through plain glass without the use of filters.

In the transmission mode the photocell and light source are located on opposite sides of the coated material and the photocell receives the light transmitted by the substrate. In the reflection mode the light beam strikes the substrate at normal incidence and a reflection is passed back along the original path until it is separated from the incident beam by a beam splitting mirror which directs the reflected beam to the photocell located on the side of the photometer head.

Radiation methods

The development of radiation-based measuring techniques has opened up new areas of coating thickness measurement. The techniques are very suitable for material combinations in which the coating and the base material are of a similar nature; the difference in atomic number should be at least 3 and at higher atomic numbers, say above 40, should be more. Measurement can be made without contact and can apply to a wide range of thicknesses with high accuracy. The source of radiation must be selected according to the nature of the measurement requirement; α-, β-, γ- and X-rays may be used. The useful life of the radioactive sources, of which some 200 different types are available, corresponds more-or-less with the half-life period. As radiation detectors, ionisation chambers, scintillation counters or counter tubes may be used. Most of the commercially available instruments, however, are equipped with so-called beta radiators and Geiger-Müller counter tubes as detectors.

The more commonly encountered beta emitters are:

Radioactive Source	*Energy (Mev.)*	*Half Life in Years*
Carbon C 14	6·16	5570
Promethium Pm 147	0·22	2·5
Caesium Cs 137	0·51	30
Krypton 85	0·70	10·5
Thallium Tl 204	0·77	3·5
Strontium Sr 90	2·18	20
Radium Ra D + E	1·17	22
Ruthenium Ru 106	3·53	1

In the radiation transmission method (Fig. 72) the radiation from a radioactive isotope is measured with a radiation meter after transmission through the material whose thickness is to be determined. The absorption by the specimen serves as a measure of the weight per unit area, which is a function of the thickness.

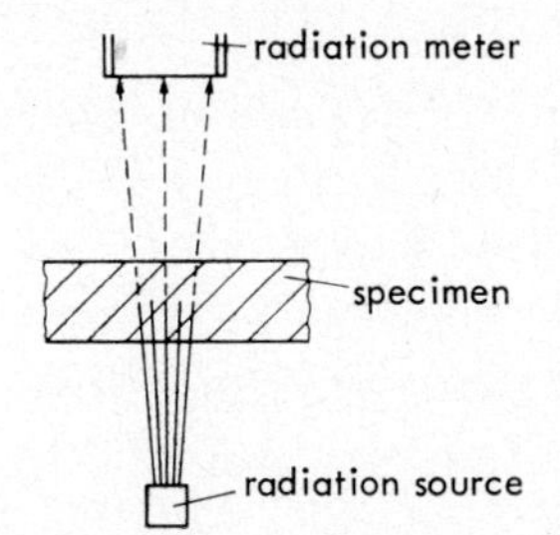

Fig. 72. Principle of radiation transmission method.

For coating thickness measurements where the coating and substrate have similar atomic numbers or densities, the differential beta transmission methods are used. Typical applications are wax and plastic foils applied to paper and aluminium sheet foil. The thickness (weight per unit area) is monitored by measuring first the thickness of the base material before the coating is applied using one measuring head, then the total thickness of coating and base by a separate measuring head. Reading A is subtracted from Reading B and displayed on a meter as the coating thickness. A typical installation for the measurement of coated sheet material is shown in Fig. 73.

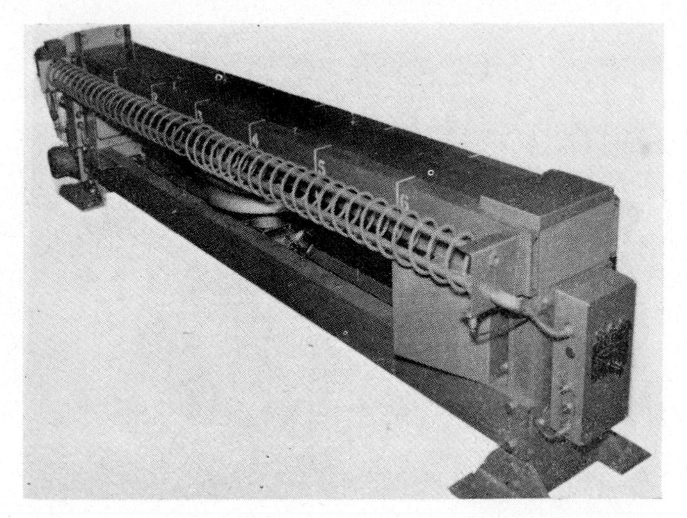

Fig. 73. The Atomat differential beta-transmission gauge of Nuclear Enterprises Ltd.
For further details see Table 6, p. 96, No. 3.

Some measurement problems are better solved by the beta backscattering method (Fig. 74). Beta rays from a radioactive isotope falling on a body are partly absorbed and partly back scattered. The intensity of the back-scatter radiation depends on the atomic number of the material concerned. If the scattering body has a coating of a second material, then the back-scatter intensity will vary with the thickness of the coating. The thickness of the coating is then determined by the saturation thickness for the beta rays. Although the transmission method determines the thickness over an appreciable area, it is possible with the back-scatter method to apply procedures which enable point measurements to be made.

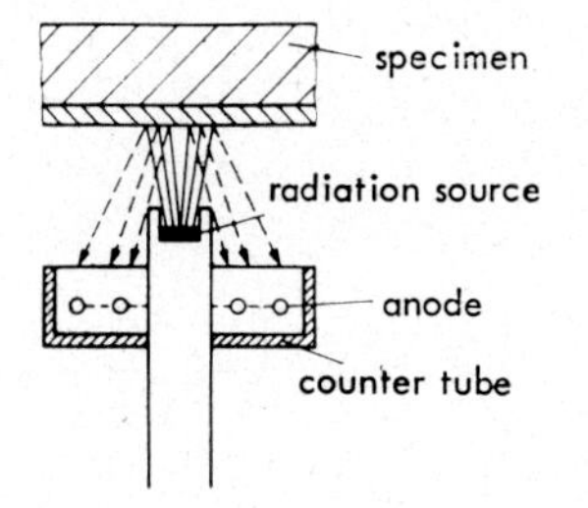

Fig. 74. Principle of beta back-scattering method.

Frieseke & Hoepfner GmbH describe as Measurement and Control System FH 46[32–34] an instrument which permits the continuous thickness control of plated and other coatings on the basis of the absorption of radioactivity (see Fig. 75). The measuring chamber (left hand part of photograph) lies above the path of the moving sheet to be controlled. The holder contains an unplated and a plated control calibration pair of sheets. The control apparatus (right hand photograph) contains as indicator instrument a recorder which registers the coating density in g/cm^2.

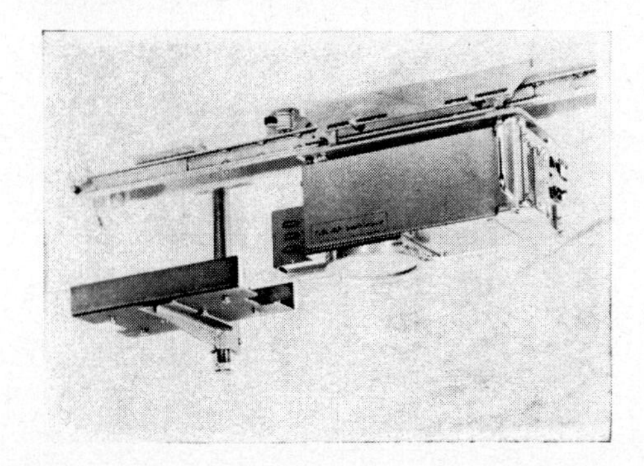

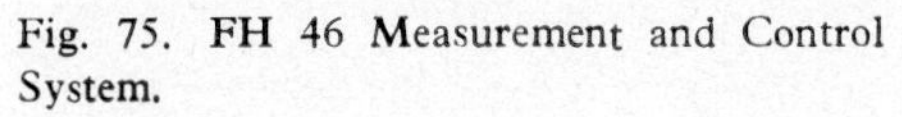
Fig. 75. FH 46 Measurement and Control System.

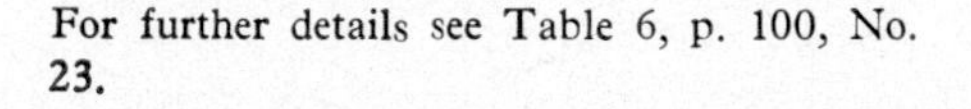
For further details see Table 6, p. 100, No. 23.

---

32. Herlitze, K. Thickness and density measurement with radioactive rays. *VDI-Z,* 1953, **103**, No. 23, 1154-1162.
33. Langel, H. The measurement of thickness of electrolytic coatings on sheets by means of nuclear radiation. *Galvanotechnik,* 1964, **55**, No. 5, 289-292.
34. Langel, H. Surface density control by application of radio isotopes. *Kerntechnik,* 1963, **5**, No. 10, 411-417.

The Atomat Beta Backscatter Gauge manufactured by Nuclear Enterprises Ltd. is used to measure the coating thickness when the process is such that the material is only accessible from one side. The measuring head is mounted on a frame which traverses the coating to be measured. This arrangement permits the measuring head to be scanned across the whole width of the coating. (Fig. 76).

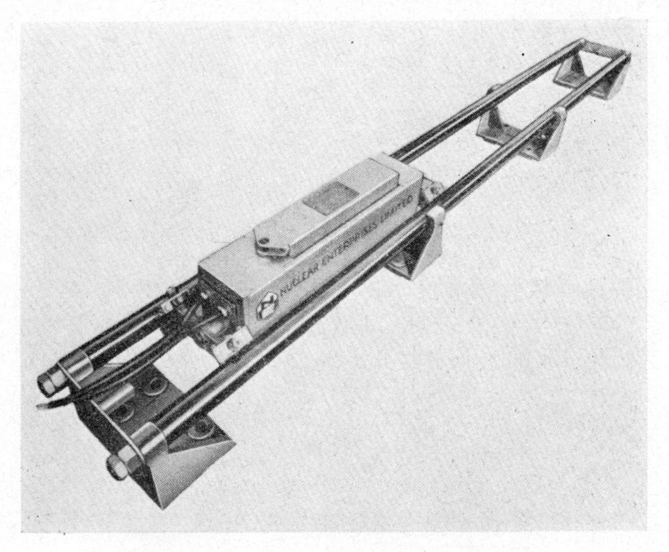

Fig. 76. Atomat Beta Backscatter Gauge.
For further details see Table 6, p. 96, No. 3.

Amongst the special coating thickness measuring intruments operating on the beta back-scatter principle are the Betascope, the Micro-Derm, the Betameter, the Burndept BN 119 and the Beta 750 Gauge.

The Betascope (Fig. 77) is supplied in the U.S.A. and the Far East by Twin City Testing Corp., and in Germany and the U.K. by Fischer. It is based on an original development by the Boeing Company. The instrument is available for all usual mains voltages, and four different models cover the requirements of commercial and scientific materials testing. These models differ mainly in the method used for counting the back-scattered beta particles. Model NX-500 has a four-figure digital indication; the evaluation of the digital indication is made using an easily generated calibration curve which, on semi-logarithmic paper, takes a straight line form. The advantages of the digital presentation are the absence of reading errors and the possibility of a rapid graphical estimation of accuracy.

The second model, AN-600, gives an analogue presentation. For direct reading of the measured coating thickness there is a clearly visible pointer

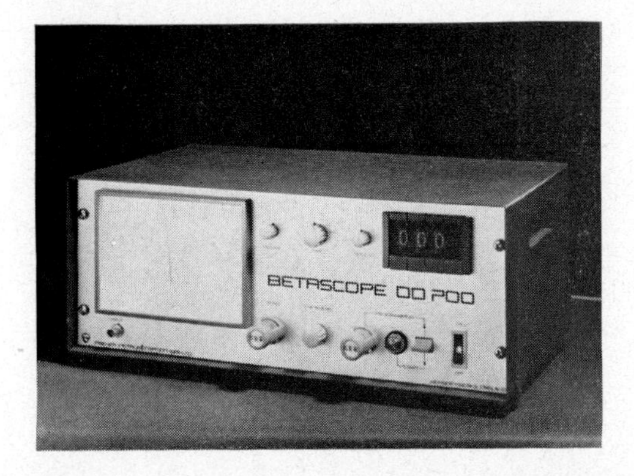

Fig. 77. Betascope DD 700

For further details see Table 6, p. 96, No. 5.

indicator on the front panel of the instrument. Compared with the digital presentation this offers a considerable reduction in time of each measurement. Model DD-700 on the other hand is provided with both digital and analogue presentation; thus the advantage of either type is available at will, while Model DZ 800 gives direct digital readout of thickness or backscatter. In general, backscatter readout is preferred by research and development staff, whereas production and inspection personnel prefer direct thickness readout.

All direct reading instruments are very quickly convertible from English to metric units and vice versa, (conversion being effected by a single adjustment of a control knob). They also have a built-in computer which automatically converts beta backscatter to the thickness units selected for display on the readout system plus a new timing concept which maintains the setting of the instrument despite any change in timing due to measurement application.

NX500 has a built-in drift compensator and a new platen system incorporating semi-precious stones into which the aperature of very precise dimensions is cut by laser. New micro-miniature energy sources have been developed to enable coated areas down to 0·4 mm diameter to be measured with high accuracy (the cross-sectional wire ends of transistor posts for example). A composite measuring table with built-in ' gun-sight ' type illuminated cross-wire system has been developed for location purposes together with a new device for measuring the through-hole plating of printed circuits, especially developed for integration into the composite measuring table. A standardised system of electronic construction facilitates production and servicing. This system, consisting of integrated circuit ' flat packs ' mounted on gold plated printed circuit cards, allows maximum utilisation of standardised assemblies. The new miniature Geiger-Müller tube/energy source system allows measurements to be made on the insides of small bore tubes.

The hand-held probe (Fig. 78) is for special requirements such as printed circuits and surfaces of large area.

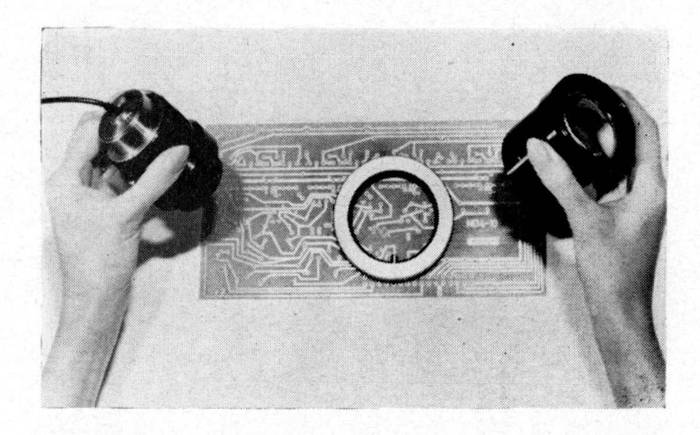

Fig. 78. Hand-held probe Z 2 NG for thickness measurement by Betascope of the conductor path of a printed circuit.

The Micro-Derm (Fig. 79) of Unit Process Assemblies Inc. is basically similar to the Betascope, differences being in the nature of the technical design. The Micro-Derm also has an analogue presentation, i.e. the coating thickness is displayed directly on a pointer instrument. The measuring head can be selected from several models, and the minimum surface provided for is 0·06 mm^2.

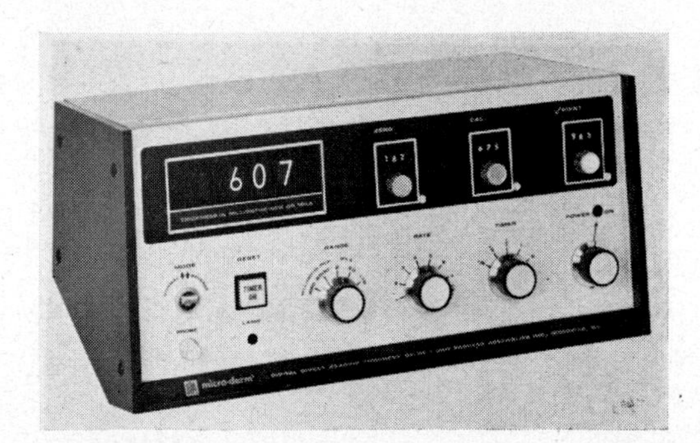

Fig. 79. Micro-Derm MD-3.

For further details see Table 6, p. 104, No. 38.

Measuring head PS 6, one of those available, is similar in its function to that of the Betascope but the radiator holder is inserted from below through an aperture in the Geiger-Müller counter tube as in Fig. 74. The separation of the source and the surface to be measured can be adjusted to steps of 0·8 mm; it needs to be adjusted to suit the aperture in use, which in turn determines the area measured. The choice of aperture depends upon the shape of the object being measured.

Instead of the measuring table, the miniature system HH 3 can be used as a hand probe; this contains the source and counter. In combination with the new measuring system carrier CB-3 (Fig. 80) it is easily possible to make measurements on objects which it would be very difficult to measure on the measuring table.

The through-hole measuring system, guide THG-1 and probe TH-1, are used to measure coatings on the walls of plated holes in printed circuit boards. By appropriate choice of isotope, the copper thickness before over-plating can be measured as well as the gold or solder over-plate thickness.

Fig. 80. Probe System CB-3 for the measurement of thickness of coatings applied to printed circuit boards.

The Betameter (Fig. 81) manufactured by EFCO Ltd. is a four figure digital instrument of British manufacture basically similar to the Betascope NX5. The latest techniques in integrated circuits are used in its construction. The special measuring table (Fig. 82) is provided with a rotating platen having

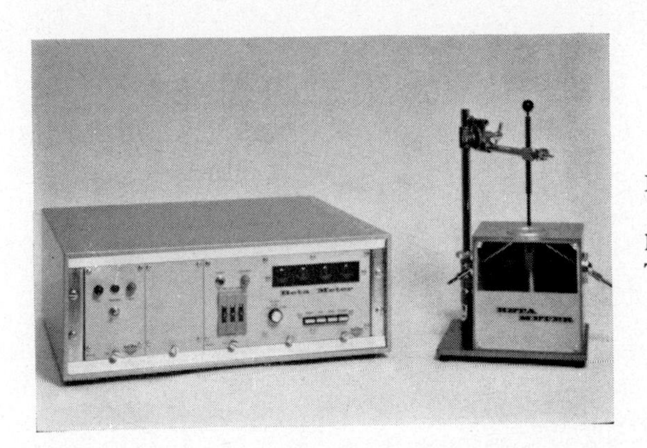

Fig. 81. Betameter.

For further details see Table 6, p. 96, No. 4.

Fig. 82. Special measuring table for the Betameter.

three circular and four slotted apertures; for special occasions a selection of small insert platens may be used. The unique design of the radio active sources and table permit two sources to be fitted simultaneously, one source retracted into the body of the source holder, the other extended under the aperture for use. The portable probe and stand are intended for use where very accurate location is required as may be encountered in the printed circuit industry. The special stand for the portable probe is designed for the coating measurement on printed circuit edge connectors or other items requiring precise location.

The Beta 750 Plating Thickness Gauge (Fig. 83) is an English development supplied by Johnson, Matthey Metals Ltd. It differs only slightly from the beta back-scatter instruments already described.

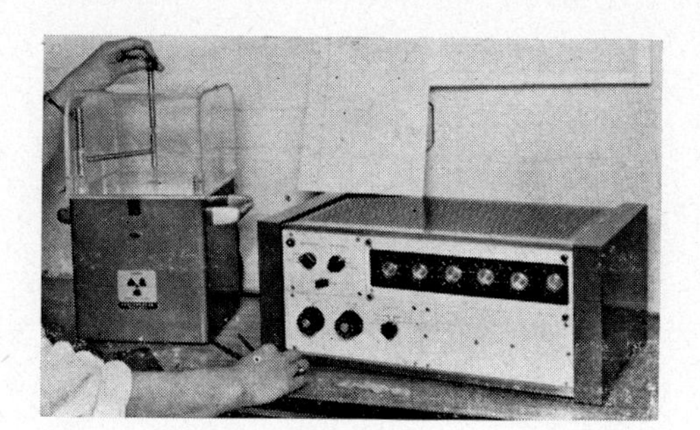

Fig. 83. Beta 750 Plating Thickness Gauge.

For further details see Table 6, p. 98, No. 6.

5 μm of gold on copper can be measured to within ± 2% accuracy in 20 seconds using a $\frac{3}{8}$ inch diameter aperture. Measurement area is 1·3 - 0·00375 cm^2. Mains operation is normal.

The Burndept BN 119 Non-Destructive Thickness Gauge* of Burndept Ltd. was developed (in conjunction with British Insulated Callender's Cables Ltd.) primarily for checking the thickness of zinc coatings on structural steel, etc., but is also being used to measure cadmium and tin on steel, silver and gold on copper, rhodium on gold and platinum on titanium. It is also suitable for metallised paint films. Being a portable instrument it is battery operated. Thickness range is 25-175 μm with an accuracy of ± 10%. A measurement takes 1 minute and a comparison test 5 seconds; as a safety measure a built-in timer prevents the meter from registering before the expiration of the 1 minute cycle. For checking the coating on galvanised iron and steel the direct reading meter can be calibrated in oz/ft^2, the range being $\frac{1}{2}$-4 oz/ft^2.

The probe head contains a centrally mounted detector and 2 Sr 90 sources. It includes sufficient shielding to make it safe for continuous use (40 hours per week) by one operator, a spring-loaded safety shutter mechanism automatically shielding the source when the probe is removed from the measuring surface.

The schematic diagram of Fig. 84 has been taken from descriptions of the Micro-Derm showing measureable material combinations; this is equally valid for all the instruments. The slope and exact shape of the curve depend

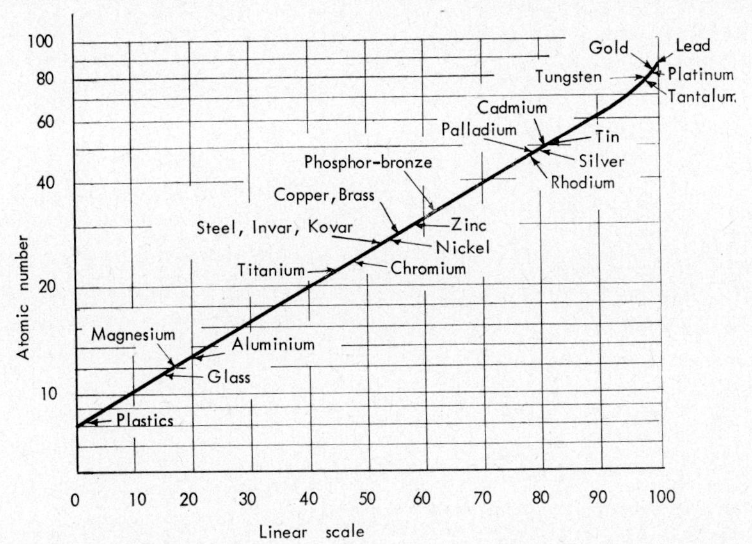

Fig. 84. Material combinations which can be measured by the beta back-scatter method.

* Stop Press: no longer available.

upon the nature of the radiation source used. In using the diagram, it can be presumed that satisfactory measurement is possible for any material combination which shows on the horizontal scale an algebraic difference of at least 15 units. Thus, gold with atomic number 79 on silver with atomic number 47 would be measured easily. Nickel, on the other hand, with atomic number 28 on iron with atomic number 26 would not be satisfactorily measured.

X-ray methods

For the non-destructive measurement of very thin coating thicknesses (well below 1 μm) X-ray techniques can be applied. As with all radiation methods, the measurement can be made on the minimum surface area. X-rays are by nature nothing more than radiation with a very short wavelength; they are not deflected by either magnetic or electric fields. A deflection can only be produced by exceedingly small lattices such as those of the crystal planes of metals. Certain materials are excited to fluorescence by X-rays; gases may be ionised.

According to the nature of the coating and base materials, the different physical properties of the X-rays may be applied to thickness measurement by different methods. For example, the determination may be made by exciting the coating or the base material to emit its charactistic radiation or to release secondary electrons. In the first case the absorption of the charactistic radiation, and in the second case the emission of secondary electrons, are in direct relationship with the coating thickness.

X-ray thickness determination has been the subject of many publications[35, 36]. Eberspächer[37] has described a measurement in terms of the angular reduction of intensity and Bierwirth[38] one using an X-ray interference technique.

The practical importance of the X-ray methods of thickness measurement, which cannot be regarded as in widespread use, is partly controversial, for the necessary expenditure is very considerable. One of the few commercially available X-ray thickness meters is the Fluoroscopy Gauge (p. 100) of EKCO Instruments Ltd. Radiation is supplied by a promethium source. The

---

35. Liebhafsky, H. and Zemany, P. Film thickness by X-ray emission spectrography. *Analytical Chemistry,* 1956, **28**, No. 7, 455-459.
36. Hess, B. Coating thickness measurement with X-rays. *Z. Angew. Phys.,* 1954, **6**, No. 1, 19-22.
37. Eberspächer, O. Non-destructive X-ray determination of the thickness of plane parallel surface coatings. *Zeitschrift für Metallkunde,* 1958, **49**, 495-498.
38. Bierwirth, G. X-ray operational control of soft nitrided components of C15 steel. *Siemens-Zeitschrift,* 1958, **32**, 365-371.

maximum attainable measuring accuracy is about ± 1%.

More significant is the automatic measurement and control of industrial (strip) plating lines. Philips Electronic Instruments supplies installations using X-ray fluorescence. With these, double-sided coatings on steel plate can be controlled to an accuracy of ± 2%. The Norelco Zinc/Aluminium Surface Density Meter (Fig. 85) measures both coatings simultaneously with a measuring head on each side of the sheet. The measurements are made without contact; each measuring head contains a water-cooled X-ray tube and its associated counter. The Norelco Tin Surface Density Meter (p. 106) is suitable for single sided or double sided measurements.

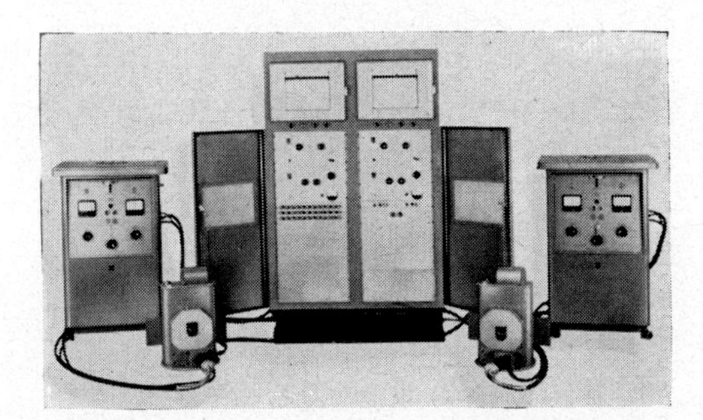

Fig. 85. Norelco Zinc/Aluminium Surface Density Meter.

For further details see Table 6, p. 106, No. 52.

The Atomat Coating and Backing X-Ray Fluorescence Gauges manufactured by Nuclear Enterprises Ltd. are suitable for the continuous measurement of tin, zinc, aluminium and chromium on sheet steel or titanium coatings applied to paper or plastic produced in sheet form. The principles are illustrated in Fig. 86 while Fig. 87 shows a typical installation.

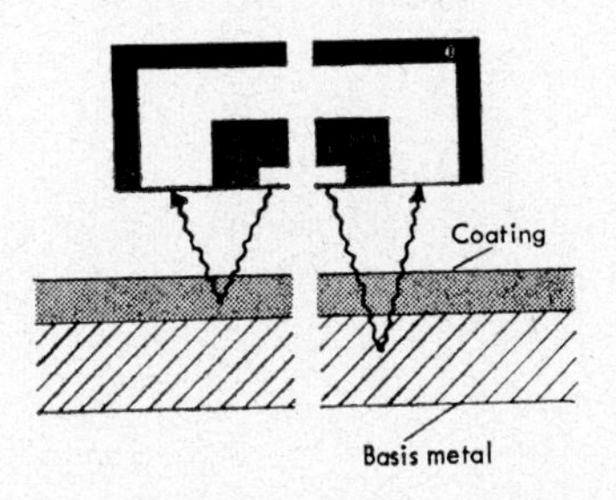

Fig. 86. X-ray fluorescence gauges employ radioisotope sources to excite characteristic radiation in the coating or backing. The diagram on the left illustrates the principle of the Atomat Coating-Fluorescence Gauge, which monitors the increase in intensity of an X-ray excited in the coating as the coating thickness is increased. The diagram on the right illustrates the principle of the Atomat Backing-Fluorescence Gauge which monitors the decrease in intensity of an X-ray excited in the backing material as the coating thickness increases.

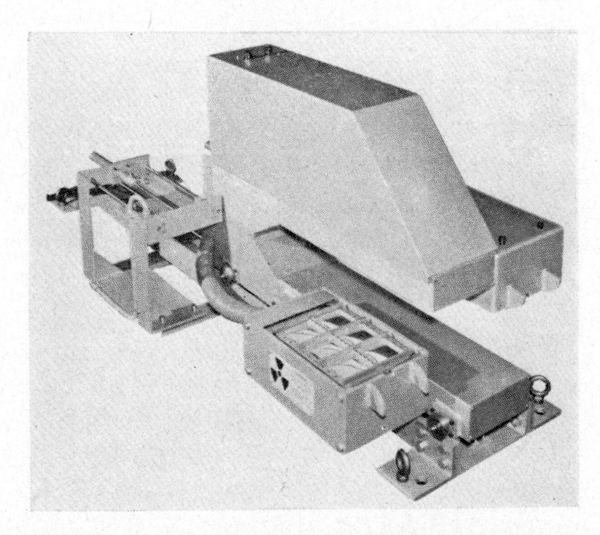

Fig. 87. Atomat X-ray fluorescence gauge for measuring tin coatings.
For further details see Table 6, p. 96, No. 3.

Crystal oscillation

Thickness measuring instruments based on the oscillation of a quartz crystal are used for one special application only, namely, to monitor the rate of deposition of vacuum evaporated films. They cannot be used to measure the thickness of an existing film and are only mentioned here for the sake of completeness. The Edwards Film Thickness Monitor consists essentially of an oscillating quartz reference crystal which is compared electronically with a similar oscillating crystal mounted in the chamber of the vacuum coating unit. As evaporation proceeds, material condenses on the chamber crystal, damping the oscillation and changing its frequency. This change in frequency is related to the film thickness, and similarly rate of change of frequency is related to the rate of film growth.

Other examples are the Deposit Thickness Monitor of Sloan Instruments Corp., the Quartz Oscillator Thickness Monitor QC1 of Klaus Schaefer GmbH and the Quartz Oscillator Thickness Monitor QSG 101 of Balzers A.-G.

Micro-resistance

The Caviderm, manufactured by Unit Process Assemblies Inc., measures the thickness and integrity of copper in plated-through holes of printed circuit boards. The measurement is made by determining the resistance of the hole using a 4-point resistance measurement, with the Caviderm readout directly in microhms (either meter or digital readout). Copper thickness is determined from a calibration curve. The minimum hole size measureable with the standard unit is 0·25 mm (0·010 inch) with special accessories available for measuring in even smaller holes. The Caviderm is able to measure copper thickness after solder is applied to the printed circuit board.

PART 3

TABLES

Table 1. Chemical Solution of Coatings : Solutions and Operating Conditions[39–43]

NOTE: 1. For details of methods and apparatus see p. 11
2. Reactions which result in noxious fumes should be conducted under a fume hood
3. Proprietary solvents are not included in this table

Coating	*Basis metal*	*Solvent*		*Temp. °C.*	*Container*
Aluminium	Iron	Sodium hydroxide	100 g/l	50-80	Steel
Brass	Steel	Chromium trioxide Sulphuric acid d 1.83	500 g/l 5 g/l	50	Lead or Stoneware
Brass	Steel	Sodium cyanide Ammonium persulphate Ammonia	80 g/l 50 g/l 20 g/l	65	Steel
Cadmium	Aluminium	Nitric acid d 1.40		20	Plastic or Stoneware
Cadmium	Steel	Ammonium nitrate	120 g/l	20	Plastic
Cadmium	Steel	Ammonium persulphate Ammonia	50 g/l 100 m/l	20	Plastic

39. Kutzelnigg, A. The testing of metallic coatings. Robert Draper Ltd., Teddington, 1963.
40. Springer, R. The removal of coatings. *Techn. Rdsch. Bern,* 1959, **24**, 22-30.
41. Straschill, M. Modern practice in the pickling of metals. Robert Draper Ltd., Teddington, 1964.
42. Weiner, R. Coating thickness measurement of electrolytic deposits. *Techn. Rdsch. Bern,* 1959, **24**, 55-61.
43. Removal of deposits: formulations, Jahrb. d Oberflächentechnik, Metall Verlag GmbH, Berlin, 1960.

TABLE 1. CHEMICAL SOLUTION OF COATINGS

Coating	*Basis metal*	*Solvent*		*Temp. °C.*	*Container*
Cadmium	Steel	Hydrochloric acid d 1.19	1 litre	20	Plastic or Stoneware
		Antimony (III) oxide	15 g		
		Water	100 ml		
Chromate	Aluminium, Zinc	Phosphoric acid d 1.75	35 ml/l		
		Chromium trioxide or	20 g/l		
		Sulphuric acid d 1.84		70-80	
		Chromium trioxide	100 g/l		
Chromate	Zinc	Glacial acetic acid	5%	20	Plastic
Chromium	Aluminium	Nitric acid d 1.51		20	Plastic
		Potassium chlorate	10 g/l		
Chromium	Copper, Magnesium,	Hydrochloric acid d 1.19	1 pt.	20	Plastic or Stoneware
	Brass, Nickel,	Water	1 pt.		
	Steel (except cast)	Commercial pickling solution	33 ml/l		
Copper	Aluminium	Nitric acid d 1.42		20	Plastic or Stoneware
Copper	Aluminium, Nickel,	Iron(III)-chloride . $6H_20$	300 g/l		
	Steel, Zinc	Copper sulphate . $5H_20$	100 g/l		
Copper	Gold	Ammonia d 0.92	1 pt.	20	Ebonite, Plastic
		Ammonium chloride	1 pt.		or Stoneware
		(20% solution)			
Copper	Gold	Nitric acid d 1.42	1 pt.	20	Plastic or Stoneware
		Water	5 pt.		

Table 1. Chemical Solution of Coatings (continued)

Coating	*Basis metal*	*Solvent*		*Temp. °C.*	*Container*
Copper	Magnesium, Zinc	Sodium hydroxide Sulphur (boil solution first for 30 min.)	100 g/l 150 g/l	85-95	Steel
Copper	non-ferrous	Sodium sulphide	100-200 g/l	20	Plastic or Stoneware
Copper	Steel	Chromium trioxide Sulphuric acid d 1.84	300-400 g/l 20 - 50 g/l	20	Lead
Copper	Steel	Sodium or potassium cyanide	20-100 g/l	60	Steel
Copper-Nickel	Steel, Zinc	Iron(III)-chloride . $6H_2O$ Copper sulphate . $5H_2O$	300 g/l 100 g/l	20	
Gold	Copper, Brass	Sulphuric acid d 1.83 Hydrochloric acid d 1.18	1000 g 250 g	60-70	Glass or Stoneware
Gold	Copper, Brass Nickel, Steel	Sodium cyanide Hydrogen peroxide (30% wt.)	125 g/l 30 ml	20-30	Plastic, Steel or Stoneware
Lead	Copper, Brass	*m*-Nitrobenzoic acid Potassium hydroxide Potassium sodium tartrate Water	34 g 50 g 68 g 1 litre		
Lead	Steel	Acetic acid (glacial) Hydrogen peroxide (30% wt.)	330 ml/l 50 ml/l	20	Ceramic, Plastic or Acid-resistant steel
Lead	Steel	Nitric acid d 1.42	20%	65	

TABLE 1. CHEMICAL SOLUTION OF COATINGS

Coating	*Basis metal*	*Solvent*		*Temp. °C.*	*Container*
Manganese phosphate	Steel	Chromium trioxide	200 g/l	20	
Nickel	Aluminium	Nitric acid d 1.42		20	Plastic or Stoneware
Nickel	Brass	Iron(III)-chloride . $6H_20$	300 g/l	20	
		Copper sulphate . $5H_20$	100 g/l		
Nickel	Copper	Sulphuric acid d 1.83	1 pt.	20	Plastic
	Brass	Nitric acid d 1.42	1 pt.		
Nickel	Steel	*m*-Nitrobenzoic acid	34 g		
		Potassium hydroxide	50 g		
		Potassium sodium tartrate	68 g		
		Water	1 litre		
Nickel	Steel	Sodium hydroxide solution	5%	20-40	Steel
		Sodium peroxide	2.5 g/l		
Nickel	Steel	Nitric acid d 1.51		20	Plastic or Stoneware
Nickel	Steel	Silver nitrate	30% wt.	20-40	Plastic
Nickel-Copper-Nickel	Steel	Iron(III)-chloride . $6H_20$	300 g/l	20	
		Copper sulphate . $5H_20$	100 g/l		
Oxide	Aluminium	Phosphoric acid d 1.75	35 ml	90	Acid resistant steel
		Chromium trioxide	20 g		
		Water	965 ml		

Table 1. Chemical Solution of Coatings (continued)

Coating	*Basis metal*	*Solvent*		*Temp. °C.*	*Container*
Oxide	Aluminium	Phosphoric acid d 1.75 Chromium trioxide addition of wetting agent desirable	30 ml 20 g/l	80-100	Acid resistant steel
Oxide	Magnesium	Chromium trioxide	150 g/l	50-70	Lead or Acid resistant Steel
Silver	Aluminium	Nitric acid d 1.42		20	Plastic or Stoneware
Silver	Aluminium Copper, Brass, Steel, Zinc, etc.	Potassium iodide Iodine	250 g/l 7.44 g/l	20	Plastic
Silver	Copper, Brass	Sulphuric acid d 1.83 Nitric acid d 1.42	250 ml 25 ml	80	Stoneware
Silver	Copper, Nickel, Steel	Sulphuric acid d 1.82 Sodium nitrate	50 g 1000 g	35-60	Enamel, Plastic or Stoneware
Silver	Copper, Nickel, Steel	Sulphuric acid d 1.83 Nitric acid d 1.37	1000 g 75 g	35-60	Enamel, Plastic or Stoneware
Tin	Aluminium, Copper, Steel, Zinc, etc.	Trichloroacetic acid	100 g/l	20	Plastic
Tin	Lead	Ammonium sulphide, yellow		20	Plastic or Stoneware

TABLE 1. CHEMICAL SOLUTION OF COATINGS

Coating	Basis metal	Solvent		Temp. °C.	Container
Tin	Bronze, Copper, Brass	Iron(III)-chloride Copper sulphate Acetic acid (56% wt.) Hydrogen peroxide (30% wt.)	75-105 g/l 135-155 g/l 310-462 ml/l as required	20	Plastic or Stoneware
Tin	Brass	Hydrochloric acid d 1.19 Water	1 pt. 1 pt.	60-70	Plastic or Porcelain
Tin	Steel	Hydrochloric acid d 1.19 Antimony oxide Water	1 litre 15 g 100 ml	20	Plastic or Stoneware
Zinc	Aluminium	Nitric acid d 1·42		20	Plastic or Stoneware
Zinc	Copper, Brass, Steel	Hydrochloric acid d 1.19 Water	1 pt. 1 pt.	20	Plastic or Stoneware
Zinc	Copper, Steel	Ammonium nitrate Hydrochloric acid, 1 N	70 g/l 70 m/l	20	
Zinc	Steel	Hydrochloric acid d 1.19 Antimony(III)-chloride or Antimony(III)-oxide Make up with water to	500 ml 3·2 g 2 g 1 litre		
Zinc	Steel	Sodium hydroxide	100-200 g/l	boil	Steel
Zinc	Steel	Sulphuric acid d 1.83 Chromium trioxide	200 g/l 15 g/l	20	Lead or Plastic
Zinc phosphate	Steel	Ammonium hydroxide d 0.91		20	
Zinc phosphate	Zinc	Chromium trioxide	50 g/l	75	

Table 2. Electrolytic Solution of Coatings :

NOTE: Reactions which result in noxious fu
For details of processes and appara
Proprietary solvents are not included

Coating	*Basis Metal*	*Solvent*	
A. ANODIC SOLUTION			
Brass	Aluminium or Steel	Ammonium nitrate	850 g/l
Brass	Steel	Ammonium nitrate Ethyl alcohol	175 g/l 175 g/l
Brass	Steel	Sodium or Potassium cyanide	100 g/l
Brass or Bronze	Steel	Sodium cyanide Sodium hydroxide	75 g/l 13 g/l
Cadmium	Aluminium, Brass, Copper, Nickel or Steel	Potassium iodide	200-400 g/l
Cadmium	Brass, Copper or Steel	Ammonium nitrate	100 g/l
Cadmium	Copper or Steel	Sodium cyanide Sodium hydroxide	100-200 g/l 30 g/l
Cadmium	Steel	Sodium cyanide	100 g/l
Chromium	Aluminium, Nickel or Steel	Sodium sulphate	100 g/l
Chromium	Aluminium, Nickel or Steel	Sulphuric acid d 1·83	70 ml/l
Chromium	Aluminium or Zinc	Sulphuric acid d 1·83 Glycerin or Copper sulphate	100 g/l 50 g/l 20 g/l
Chromium	Brass, Copper, Nickel silver, Magnesium or Steel	Sodium hydroxide	100-150 g/l
Chromium	Brass	Sulphuric acid d 1·83 Glycerin	880 ml/l 7 ml/l

39. Kutzelnigg, A. The testing of metallic coatings. Robert Draper Ltd., Teddington, 1963.
40. Springer, R. The removal of coatings. *Techn. Rdsch. Bern,* 1959, **24**, 22-30.
41. Straschill, M. Modern practice in the pickling of metals. Robert Draper Ltd., Teddington, 1964.

Solutions and Operating Conditions[39–44]

mes should be conducted under a fume hood.
tus see p. 17.
in this table.

Temp. °C.	Current density A/dm²	Voltage V	Cathode	Contamer
20	3		Steel	Plastic or Stoneware
20	4		Steel	Plastic or Stoneware
20	0·5-1		Steel	Plastic, Steel or Stoneware
20	3		Copper	Plastic or Stoneware
20	3			Plastic or Stoneware
20		4-10	Steel	Plastic, Steel or Stoneware
20	2-5		Steel	Plastic, Steel or Stoneware
20		4-10	Steel	Plastic, Steel or Stoneware
20	6		Lead	Lead, Plastic or Stoneware
20	6		Lead	Lead, Plastic or Stoneware
20	5-10	5-10	Lead	Lead, Plastic or Stoneware
25-35	2-5	2-4	Steel	Plastic, Steel or Stoneware
20		6	Lead	Lead or Stoneware

42. Weiner, R. Coating thickness measurement of electrolytic deposits. *Techn. Rdsch. Bern,* 1959, **24**, 55-61.
43. Removal of deposits: formulations, Jahrb. d Oberflächentechnik, Metall Verlag GmbH, Berlin, 1960.
44. Pollak, A. The electrolytic stripping of electrolytic deposits. *Metall-oberfl.,* 1951, **3**, No. 1, 8-9.

Coating	*Basis Metal*	*Solvent*	
Chromium	Nickel or Steel	Sodium hydroxide	50-100 g/l
Chromium- Nickel- Copper	Aluminium or Steel	Phosphoric acid d 1·70 Triethanolamine	750 g/l 250 g/l
Copper	Aluminium	Nitric acid d 1·42	5%
Copper	Aluminium, Nickel or Steel	Ammonium nitrate	850 g/l
Copper	Aluminium, Nickel or Steel	Ammonium nitrate Sodium-potassium tartrate	100 g/l 80 g/l
Copper	Steel	Ammonium nitrate Ethyl alcohol	175 g/l 175 g/l
Copper	Steel	Chromium trioxide Sulphuric acid d 1·83	100-300 g/l 10-50 g/l
Copper	Steel	Potassium or Sodium cyanide Sodium hydroxide up to pH 13·5	100-200 g/l
Copper	Steel	Copper cyanide Sodium cyanide Sodium phosphate . 12 H_2O	8-80 g/l 12-120 g/l 60 g/l
Copper	Steel	Sodium cyanide Sodium hydroxide Hydrogen peroxide	100 g/l 10 g/l 1 ml/l
Copper	Steel	Sodium nitrate	150-200 g/l
Copper	Steel	Sodium nitrite	250 g/l
Copper	Zinc	Sodium sulphide	120 g/l
Copper	Zinc Zinc diecastings	Potassium iodide Sodium phosphate . 12 H_2O	880 g/l 60 g/l
Gold	Copper	Potassium cyanide Alum Potassium cyanoferrate(II)	50 g/l 20 g/l 30 g/l
Gold	Brass, Copper or Steel	Sodium cyanide Sodium hydroxide	90 g/l 15 g/l
Gold	Silver	Hydrochloric acid d 1·19	
Gold	Steel	Potassium cyanide	100-150 g/l

Table 2. Electrolytic Solution of Coatings (continued)

Temp. °C.	*Current density* A/dm^2	*Voltage* V	*Cathode*	*Container*
20	2-5	2-4	Steel	Steel or Stoneware
65-90	10			Stoneware
20		6	Steel	Plastic
20	4		Steel	Plastic or Stoneware
20	4		Steel	Plastic or Stoneware
20	4		Steel	Plastic or Stoneware
20	1-10	4-6	Lead	Lead lining
20	0·5-5	4-6	Steel	Plastic or Steel
40-90	1-15	2-5		
50-60	1	2-4	Steel	Steel
20	2-5	4-6	Steel	Plastic or Steel
20	1-5	4-6	Steel	Plastic, Steel or Stoneware
20	2	4	Steel	Plastic or Stoneware
20	3			Plastic or Stoneware
		5-10	Copper	
20		6-8	Steel	Plastic, Steel or Stoneware
20		3-4	Carbon	Plastic
30	4-8	4-6	Copper	

Coating	*Basis Metal*	*Solvent*	
Lead	Copper, Brass or Steel	Ammonium acetate Sodium acetate	200 g/l 200 g/l
Lead	Steel	Sodium hydroxide Metasilicate Pot. Sod. tartrate	100 g/l 75 g/l 50 g/l
Lead	Steel	Sodium nitrate, pH 6-10	500 g/l
Nickel	Aluminium	Sulphuric acid d 1·83	
Nickel	Aluminium, Brass, Copper or Steel	Ammonium nitrate Ammonium thiocyanate	400 g/l 45 g/l
Nickel	Aluminium, Brass, Copper or Steel	Ammonium nitrate Sodium thiocyanate	30 g/l 30 g/l
Nickel	Brass or Copper	Hydrochloric acid d 1·19	150 ml/l
Nickel	Brass, Copper or Steel	Sulphuric acid d 1·83 Glycerin or Copper sulphate	100 g/l 50 g/l 20 g/l
Nickel	Copper or Steel	Chromium trioxide Boric acid	200 g/l 30 g/l
Nickel	Magnesium	Hydrofluoric acid (35-40%) Sodium nitrate	15-20% wt. 2% wt.
Nickel	Brass or Zinc	Sodium thiocyanate Sodium bisulphite	100 g/l 100 g/l
Nickel	Steel	Sodium nitrate	300-500 g/l
Nickel	Zinc	Sulphuric acid d 1·83	70%
Silver	Aluminium, Nickel or Steel	Ammonium nitrate	860 g/l
Silver	Brass or Copper	Potassium fluoride	100 g/l
Silver	Brass, Copper or Nickel silver	Potassium thiocyanate	180 g/l
Silver	Brass, Copper, Nickel silver, Nickel or Zinc	Sulphuric acid d 1·83 Nitric acid d 1·37	1000 g 75 g
Silver	Nickel or Steel	Ammonia Sodium nitrate	5 g/l 100 g/l
Silver	Steel	Sodium- or Potassium-cyanide Sodium hydroxide	50-75 g/l 20 g/l

Table 2. Electrolytic Solution of Coatings (continued)

Temp. °C.	*Current density* A/dm²	*Voltage* V	*Cathode*	*Container*
20	5		Steel	Plastic
82	2-4			
20-80	2-20		Steel	Plastic or Steel
			Lead	Lead
20	4		Steel	Plastic or Stoneware
20	4		Steel	Plastic or Stoneware
		6	Carbon	
20	4-10	4-8	Lead	Lead, Plastic or Stoneware
85	1		Lead	Lead
20	2		Carbon	Plastic
20	2	6-10	Steel	Plastic, Steel or Stoneware
90	6-10	6	Steel	Steel or Stoneware
20	2		Lead	Lead, Plastic or Stoneware
20	2-6		Steel	Plastic or Stoneware
20	2		Carbon	Plastic
20	2-5		Carbon	Plastic or Stoneware
20		2-3	Lead	Lead, Plastic or Stoneware
20	2		Steel	Plastic, Steel or Stoneware
20	1-2	4-5	Steel	Plastic, Steel or Stoneware

Coating	*Basis Metal*	*Solvent*	
Tin	Aluminium	Sulphuric acid d 1·83	70 g/l
Tin	Brass, Copper or Steel	Sodium hydroxide	100-150 g/l
Tir	Steel	Hydrochloric arid	1 N
Zinc	Aluminium, Brass, Copper or Steel	Sodium sulphate . 10 H_2O Zinc sulphate . 7 H_2O	100 g/l 20 g/l
Zinc	Aluminium, Brass, Copper, Nickel or Steel	Potassium or Sodium chloride	100 g/l
Zinc	Steel	Ammonium nitrate Grape sugar	500 g/l 5 g/l
Zinc	Steel	Sodium chloride Zinc sulphate	20% 10%
Zinc	Steel	Sodium hydroxide	100 g/l

B. CATHODIC SOLUTION

Chromating	Aluminium, Cadmium or Zinc	Sodium hydroxide Potassium or Sodium cyanide	5 g/l 50 g/l
Chromating	Aluminium, Cadmium or Zinc	Sodium hydroxide or Sodium cyanide	100 g/l 50 g/l

Table 2. Electrolytic Solution of Coatings (continued)

Temp. °C.	*Current density* A/dm^2	*Voltage* V	*Cathode*	*Container*
20	3		Lead	Plastic or Stoneware
20		6	Steel	Plastic, Steel or Stoneware
20	3		Carbon	
20	3		Lead	Lead, Plastic or Stoneware
20	3		Carbon	Plastic or Stoneware
20	3		Steel	Plastic or Stoneware
20	3		Steel	
20	2		Steel	Plastic, Steel or Stoneware
20	15		Steel	Plastic, Steel or Stoneware
20		6-8	Steel	Plastic, Steel or Stoneware

Table 3. Guide to Destructive Tests

Note : For further details of commercial instruments see Table 4.
In other cases, see the page number given here or the subject index

Generally Applicable Methods

Stripping methods using
- Analytical determination,
- Gravimetric determination, or the
- Dial (clock) gauge principle.

Sectioning methods
- Normal section (ASTM A219; DIN 50 950) (see p. 1)
- Inclined section
- Arc section (Mesle chord) method (see p. 7).

Spectrographic method.

Generally Applicable Instruments

Elmymeter
Millimess
Rossman wet and dry film thickness gauge Type 296
Platimeter
Coating thickness meter Type S 1566
Tolerator
Rossman dry film thickness gauge Type 233.

Coating	*Basis Material*	*Method*	*Instrument*
Organic Coatings			
Dry films		ASTM D 1005-51 (1966)	PEG Gage Penetrometer PIG Gauge
Wet films		ASTM D 1212-54 (1965) DIN-E 53 158	Pfund film thickness tester Rossman wet film thickness gauge Wet film thickness gauge
Anodic Oxide Coatings	Aluminium	ASTM B 137-45 (1965) DIN 50 943 DIN 50 944 DIN 50 948	Bell Laboratory Tester Gratometer
Metal Coatings			
Brass	Aluminium, Non-metals or Steel		Kocour Model G 660
Bronze	Steel		Jet test

TABLE 3. DESTRUCTIVE TESTS

Coating	*Basis Material*	*Method*	*Instrument*
Cadmium	Aluminium, Brass, Copper, Nickel, Non-metals or Steel	DIN 50 955	Coulometric Plating Gauge Coulometric Plating Thickness Tester
	Aluminium, Brass, Copper, Steel, Zinc, etc.	DIN 50 951	Kocour Model 660 Jet test
	Steel	Calorimetric process of Krijl and Melse (p. 13)	
	Steel	Drop test (ASTM A 219-58) (see p. 13)	
Chromium	Aluminium, Brass, Copper, Nickel, Nickel silver, Non-metals, Stainless steel or Steel		Coulometric Plating Gauge Coulometric Plating Thickness Tester Kocour Model 660
	Aluminium, Brass, Copper, Nickel, Non-metals or Steel	DIN 955 (see p. 17)	
	Copper, Steel, Zinc or Zinc diecastings	Spot test (ASTM A 219-58; BS 1223-1959) (see p. 15)	
	Nickel or Nickel-copper	Spot test (DIN 50 953) (see p. 15)	
Cobalt	Steel		Jet test
Copper	Aluminium, Aluminium-bronze, Brass, Nickel, Nickel-steel alloys, Non-metals, Steel, Tungsten or Zinc diecastings		Coulometric Plating Gauge Coulometric Plating Thickness Meter Kocour Model 660
	Aluminium, Nickel, Non-metals, Steel or Zinc	Anodic solution (DIN 50 955) (see p. 17)	
	Aluminium, Nickel, Steel or Zinc	DIN 90 951	Jet test
Gold		Chemical solution of basis metal (see p. 12)	

Table 3. Destructive Tests (continued)

Coating	*Basis Material*	*Method*	*Instrument*
Lead	Aluminium, Brass, Copper, Kovar, Non-metals, Silver, Steel or Zinc		Coulometric Plating Gauge Coulometric Plating Thickness Meter Kocour Model 660
	Aluminium, Brass, Copper, Nickel, Non-metals or Steel	Anodic solution (DIN 50 955) (see p. 17)	
	Copper, Silver or Steel		Jet test
	Steel	Drop test (ASTM A 219-58) (see p. 15)	
Lead-tin alloys	Aluminium, Brass, Copper, Non-metals or Steel		Kocour Model G 660
Nickel	Aluminium, Brass, Copper, Non-metals or Steel	Anodic solution (DIN 50 955) (see pp. 17 and 23)	
	Aluminium, Brass, Copper, Inconel, Kovar, Molybdenum, Non-metals, Steel or Tungsten		Kocour Model 660
	Aluminium, Brass, Copper, Steel or Zinc	DIN 50 951	Jet test
	Steel	Calorimetric process of Krijl and Melse (see p. 13).	
Platinum		Chemical solution of basis metal (see p. 12)	
Silver	Aluminium, Brass, Copper, Nickel, Nickel-silver, Non-metals, Steel or Tin		Coulometric Plating Gauge Coulometric Plating Thickness Meter Kocour Model 660
	Aluminium, Brass, Copper, Steel, Zinc	DIN 50 951	Jet test
	Aluminium, Brass, Copper, Nickel, Non-metals or Steel	Anodic solution (DIN 50 955) (see p. 17)	
	Phosphor bronze	Calorimetric process of Krijl and Melse (see p. 13).	

TABLE 3. DESTRUCTIVE TESTS

Coating	*Basis Material*	*Method*	*Instrument*
Tin	Aluminium, Brass, Cadmium, Copper, Nickel, Nickel-silver, Non-metals or Steel		Coulometric Plating Gauge Coulometric Plating Thickness Meter Kocour Model 660
	Aluminium, Brass, Copper, Nickel, Non-metals or Steel	Anodic solution (DIN 50 955) (see pp. 17 and 23)	
	Aluminium, Brass, Copper, Steel or Zinc	DIN 50 951	Jet test
	Brass or Copper	Titration with methylene blue according to Diesing and Schneider (see p. 11)	
	Brass, Copper or Steel	Drop test (ASTM A 219-58) (see p. 13)	
	Copper wire	Seddon test (see p. 19)	
	Copper wire	DIN 51 213 (see p. 11)	
	Steel	Potential-time diagram (see p. 18)	
	Steel	DIN 50 954 (see p. 11)	
Tin-zinc alloys	Brass, Copper, Non-metals or Steel		Kocour Model 660
	Steel		Jet test
Zinc	Aluminium, Brass, Copper, Nickel, Non-metals or Steel	DIN 50 955 Anodic solution	Coulometric Plating Gauge Coulometric Plating Thickness Meter Kocour Model 660
	Copper or Steel	DIN 50 951	Jet test
	Steel	Calorimetric process of Krijl and Melse (see p. 13)	
	Steel	Preece Test (see p. 15)	
	Steel	Drop test (ASTM A 219-58) (see p. 15)	
	Steel	Voltage-time diagram (DIN 50 932) (see p. 18)	
	Steel	DIN 50 952 (see p. 11)	
	Steel strip or wire		ZABA apparatus (see p. 12)
	Steel wire	DIN 51 213 (see p. 11)	

Table 4. Destructive Thic

	Name of instrument	*Manufacturer**	*Price Category***	*Principle of measurement*	*Coating*
1	Bell Laboratory Tester Fig. 24, page 24	Bell Telephone Laboratories Inc.		Breakdown voltage	Insulating
2	Coating Thickness Meter S 1566 and S 1567 Fig. 11, page 10	H. C. Kröplin GmbH	A	Clock gauge	Dry
3	Coulometric Plating Gauge Page 21	Thorn Bendix Ltd.	C	Coulometric	Metal
4	Coulometric Plating Thickness Meter Model SS Fig. 22, page 21	M. L. Alkan Ltd.	C	Coulometric	Metal
4a	Dektak Page 10	Sloan Instruments	D	Diamond stylus	Metal
5	Elmymeter Type ET	Dr.-Ing. Perthen GmbH	C	Clock gauge	Dry
6	Gratometer Fig. 24, page 24	Langbein-Pfanhauser Werke AG., Vienna	C	Breakdown voltage	Insulating
7	Jet test apparatus Fig. 15, page 14	British Drug Houses Ltd. Langbein-Pfanhauser Werke AG., Neuss Ströhlein & Co.	B	Chemical stripping	Electrodeposi
8	Kocour Model 660 Fig. 20, page 19	Kocour Company BFI-Elektronik GmbH	C	Coulometric	Metal

* Or main agents or associates. For addresses see p. 118.

**A Less than £10 ($24)

B £10-£50 ($24-$120)

C £50-£500 ($120-$1200)

D More than £500 ($1200)

These price indications are for guidance only and are likely to vary from time to tim and from place to place. They are for the basic equipment only.

† Quoted in metric units, but most instruments are also available with inch calibrations .

TABLE 4. DESTRUCTIVE INSTRUMENTS

ess Measuring Instruments

asis material	*Dimensions, Weight, Power supply*	*Instrument characteristics†*
Metal	60 Hz	Accuracy ± 10% Probe pressure 1-2 kg Probe head, sphere, 3 mm dia. Test voltage: 0-1500 V in 25 V steps
	100 g	Measuring surface 4 mm^2 Range + 200 to − 3500 μm Probe radius 0·5 mm
Metal	240 × 150 × 220 mm^3 7·3 kg Mains driven	Range 0.05-100 μm Measuring surface 7.6 mm dia. Accuracy ± 5%
Metal or non-metal	300 × 200 × 100 mm^3 4 Kg	Measuring range 0·05-50 μm Accuracy ± 3%
	15½" × 12½" × 8" 30 lb Mains	Vertical magnification 1 000 to 1 000 000 ×
	225 × 310 × 360 mm^3 5·5 kg Mains or battery	Internal and external measurement: 4 ranges ± 3-10-30-100 μm Accuracy ± 1% of range Probe pressure 50 g
Metal	250 × 200 × 140 mm^3 5·5 kg Mains driven	Measuring surface 1 mm^2 Range 2-25 μm Accuracy ± 10%
		Measuring surface about 150 mm^2 Accuracy ± 15%
Metal or non-metal	430 × 250 × 250 mm^3 13·0 kg Mains driven	Measuring surface 8mm^2 (4.5 mm^2) Range 0·07-100 μm (Cr 0·007-100 μm) Accuracy ± 5% Accessories for wires and small components

	Name of Instrument	*Manufacturer**	*Price Category***	*Principle of Measurement*	*Coating*
9	Millimess No. 1003 Fig. 8, page 8	C. Mahr	B	Clock gauge	Dry
10	Paint Inspection Gauge (PIG) Fig. 6, page 6	Elcometer Instruments Ltd. Stierand Prüfgeräte GmbH & Co. KG	B	Optical measurement of cut in coating	Dry
11	PEG Thickness Gauge Fig. 5, p. 5	Gardner Laboratories Inc. Erichsen GmbH & Co.	B	Roller wheel	Dry
12	Penetrometer Fig. 4, p. 5	Gardner Laboratories Inc. Erichsen GmbH & Co.	C	Penetration by a needle	Dry
13	Pfund film thickness meter Fig. 3, page 4	Koehler Instrument Co. Inc.	B	Penetration by a convex lens	Wet
14	Platimeter Fig. 10, page 9	Theodor Rapp K.G.	B	Clock gauge	Dry
15	Rossman Type 233 Dry Film Thickness Gauge Fig. 12, page 10	Erichsen GmbH. & Co.	A	Clock gauge	Dry
16	Rossman Type 333 Wet Film Thickness Gauge	Erichsen GmbH & Co.	A	Penetration of quadratically arranged teeth	Wet
	Elcometer Types 333, 115, 154 Figs. 2a and 2b, pp. 3 and 4	Elcometer Instruments Ltd.	A		
17	Rossman Wet and Dry Film Thickness Gauge Type 296 Fig. 13, page 10	Erichsen GmbH & Co.	A	Clock gauge	Wet and dry
18	Tolerator Fig. 9, page 9	Ernst Leitz GmbH	C	Clock gauge	Dry
19	Wet Coating Thickness Gauge Fig. 1, page 3	Gardner Laboratories Inc.	A	Penetration of an eccentric roller	Wet
		Erichsen GmbH & Co.	A		
		Elcometer Instruments Ltd.	A		
		Sheen Instruments Ltd.	A		
20	ZABA Apparatus Fig. 14, page 12	Keller & Bohacek K.G.	B	Measurement of evolved gas	Steel wire

Table 4. Destructive Thickness Measuring Instruments (continued)

Basis Material	*Dimensions, Weight, Power supply*	*Instrument characteristics*†
		Range ± 50 µm Accuracy ± 0.5 µm Probe pressure 100 g
	110 × 86 × 26 mm^3 350 g	Range 0-2000 µm Accuracy ± 2-20 µm Built-in measuring microscope of × 50 magnification
		Range 25-165 µm Scale divisions 12·7 µm
		Range 0-51 µm Accuracy ± 0·25 µm Scale divisions 1·27 µm Needle pressure 0-4·5 kg according to hardness of coating Indication of end of measurement by signal lamp
		Range 0.00225-0.36100 mm or 0.089-14.21 mils
		Range 0·5-2000 µm Accuracy ± 0·5 µm
		Range 0-1000 µm Accuracy ± 5 µm
		Range 0-120 or 0-1200 µm Accuracy ± 5 µm or ± 50 µm
		Range 0-500 µm
	480 × 250 × 320 mm^3 30 kg	Range ± 70 µm Accuracy ± 1-1·5 µm Probe pressure 200 g Suitable for internal and external measurements
		8 different ranges from 0-1500 µm Accuracy ± 1 - ± 50 µm according to range
Steel wire		Wire diameter 0·1-4, 0·1-6 or 0·1-9 mm Accuracy ± 1-2%

Table 5. Guide to Non-destructive Tests

Generally Applicable Methods

Clock gauge principle using difference measurements (see p. 27)
Gravimetric by differential weighing (see p. 27)
Optical by measurements at steps (see p. 51)

Coating	*Basis material*	*Method of measurement and Instrument*
1. Ferromagnetic	**Ferromagnetic**	Eddy current methods (see p. 42) Dermitron* Elcotector MK III* Nickelscope EN1
Nickel	Steel	Magnetic attraction methods (see p. 27) Magne-Gage Mikrotest
		Thermoelectric method (see p. 50)
2. Ferromagnetic	**Non-ferromagnetic**	Eddy current methods (see p. 42) Dermitron* Elcotector MK III* Nickelscope EN
		Magnetic attraction methods (see p. 27) Magne-Gage
		Radiation methods+ (see p. 56) Atomat beta backscatter gauge Betameter Betascope Beta 750 plating thickness gauge Burndept BN 119 Fluoroscopy gauge N 683 FH 46 Measurement and control system Micro-Derm PW 4390 coating thickness meter
		Thermoelectric method (see p. 50)
3. Non-ferromagnetic	**Ferromagnetic**	Magnetic attraction methods (see p. 27) BSA/Tinsley pencil gauge Chemigage Elcometer pull-off gauge Inspector thickness gauge Lectromag Magne-Gage MA S/P coating thickness meter Mikrotest

* Difference in conductivity between coating and base material must be adequate. See Fig. 51, page 42.
+ Difference in atomic number between coating and base material must be adequate. See Fig. 84, page 65.

TABLE 5. NON-DESTRUCTIVE TESTS

Coating	*Basis Material*	*Method of measurement and Instrument*
		Magnetic induction methods (see p. 32) Accuderm M or D Coating meter 2.099 Diameter coating thickness meter Elcometer thickness gauge 101 Elektrotest standard or FP Elmicron FE Leptoscope SMG, EP, T500 or SMG 8 Magnus Junior Mini-Leptoscope Minitector G Monimeter 2.094 Permascope ES 4 Reluctance-type thickness gage SM1, SM2 and SM3 coating thickness meters
		Radiation methods+ (see p. 56) Atomat coating and backing X-ray fluorescence gauge Betameter Betascope Beta 750 plating thickness gauge Burndept BN 119 thickness gauge FH 46 Measurement and control system Fluoroscopy gauge N 683 Micro-Derm Norelco zinc/aluminium or tin surface density meter PW 4390 coating thickness meter
		Eddy current methods (see p. 42) Dermitron* Elcometer 133 Elcotector MK III* Elmicron FE Permascope ES 4
Aluminium or Zinc		Norelco zinc/aluminium surface density meter
Insulating coatings		Capacitative method (see p. 49)
Insulating coatings (transparent)		Optical methods (see p. 51) Angstrometer HI Interference microscope Microscopes of light section and interference types Nachet 300 microscope Optical coating thickness meter

[*continued over*

Table 5. Non-destructive Tests (continued)

Coating	*Basis material*	*Method of measurement and Instrument*
Insulating coatings (transparent, coloured)		Photoelectric method (see p. 55) Modulated beam photometer
Tin	Steel	Norelco tin surface density meter
4. Non-ferromagnetic	**Non-ferromagnetic**	Eddy current methods* (see p. 42) Dermitron Elcotector MK III* Nickelscope EN 1 Permascope EC 3
		Radiation methods+ (see p. 56) Atomat coating and backing X-ray fluorescence gauge Betameter Betascope Beta 750 plating thickness gauge Burndept B119 thickness gauge FH 46 Measurement and control system Fluoroscopy gauge N 683 Micro-Derm PW 4390 coating thickness meter
Insulating coatings	Non-ferromagnetic	Eddy current methods (see p. 42) Eddy Gauge Model 133 Elmicron NF Filmeter NRL 1 Isometer 2.082 Leptometer NET Minitor thickness gauge*
		Magnetic induction methods (see p. 32) Elektrotest U
Insulating or paint coatings	Non-ferromagnetic	Angstrometer Capacitative method (see p. 49) Ultrasonoscope film thickness meter
Insulating (transparent)		Optical methods (see p. 51) Angstrometer HI interference microscope Microscopes of interference or light section types Nachet 300 microscope Optical coating thickness meter GBS 02
Insulating (transparent, coloured)		Photoelectric method (see p. 55) Modulated beam photometer
Titanium	Paper or plastic	Atomat coating and backing X-ray fluorescence gauge

Table 5. Non-destructive Tests (continued)

Coating	*Basis Material*	*Method of measurement and Instrument*
Copper	In holes of printed circuit boards	Micro-resistance Caviderm
5. Vapour deposits	—	Optical methods (see p. 51) Angstrometer Optical coating thickness meter Photoelectric method (see p. 55) Modulated beam photometer Quartz oscillator methods (see p. 68) Deposit thickness monitor DTM 3 Film thickness monitor FTM2 Quartz oscillator thickness monitor QC1 Quartz oscillator thickness monitor QSG 101

Table 6. Non-destructive Coating

	Name of Instrument	Manufacturer*	Price Category**	Principle of Measurement	Coating
1	Accuderm Models M and D Fig. 46, page 39	Unit Process Assemblies Inc.	C	Magnetic induction	Non-ferro-magnetic
2	Angstrometer Model M-100	Sloan Instruments Corp.	D	Optical interference	Transparent. Opaque with steps(45–47)
3	Atomat Figs. 73, 76, 86 and 87 Pages 58, 60, 67 and 68	Nuclear Enterprises Ltd.	D	Differential beta-transmission, Beta backscatter, X-ray fluorescence, preferential absorption of X-rays	Metallic and Non-metallic
4	Betameter Fig. 81, page 63	EFCO Ltd., Sel-Rex	D	Beta backscatter	Metal or non-metal
5	Betascope Fig. 77, page 61	Twin City Testing Corp. Helmut Fischer GmbH	D	Beta backscatter, Beta trans-mission	¶Metal or non-metal

* Or main agents or associates. For addresses and for local agents see p. 118.

**A Less than £10 ($24)
B £10-£50 ($24-$120)
C £50-500 ($120-$1200)
D More than £500 ($1200)
These price indications are for guidance only and are likely to vary from time to time and from place to place. They are for the basic equipment only.

† Quoted in metric units, but most instruments are also available with inch calibrations .

‡ Conductivity difference between coating and base must be adequate (see Fig. 51)

¶ Atomic number difference between coating and base must be adequate (see Fig. 79)

Literature references are on p. 108.

TABLE 6. NON-DESTRUCTIVE INSTRUMENTS

Thickness Measuring Instruments

Basis material	*Dimensions, Weight, Power supply*	*Instrument characteristics*†
Ferromagnetic	Mains or battery	3 ranges between 0-225 μm Presentation digital or meter
Non-transparent	254 × 250 × 375 mm^3 30 kg Mains supply	Range 0·05-20 μm Accuracy ± 0·01 μm
Metal or non-metal		Continuous measurement of sheet or strip material, accuracy ± 1% of measured value
Metal or non-metal	Mains supply	Measuring area diameter 1·2 mm-1·8 mm Measuring area slot 0·38 mm-1·6 mm Three different radioactive sources Accuracy ± 1-10% dependent on materials and source Digital readout Measuring table rotating platen Portable probe and measuring stand for printed circuits
Metal or non-metal	490 × 315 × 200 mm 12.5 kg Mains Supply Type NX500 Type AN600 Type DD700 Type DZ800	Min. measuring surface 0.13 mm^2 Min. diameter of specimen 0.3 mm Range: ca. 250Å upwards (depending upon materials and sources) Accuracy: ± 1 - 10% (depending upon materials and sources) Accessories: Standard table Z1NG; Mobile table Z2NG; Miniature probe Z4NG; Multi-purpose table Z5NG Readout facility: Type NX 500 Digital backscatter; AN 600 Direct reading analogue; DD 700 Direct reading analogue plus digital backscatter; DZ 800 Direct reading digital thickness plus backscatter

[*continued over*

	Name of Instrument	*Manufacturer**	*Price Category***	*Principle of Measurement*	*Coating*
6	Beta 750 plating thickness gauge Fig. 83, page 64	Panax Equipment Ltd.		Beta backscatter	¶Metal or non-metal
7	BSA-Tinsley pencil thickness gauge Fig. 30, page 30	Evershed & Vignoles Ltd. Detakta	B	Magnetic attraction	non-ferro-magnetic
8	Burndept BN119 non-destructive thickness gauge Page 65	Burndept Ltd.	No longer available	Beta backscatter	Magnetic or non-magnetic
8a	Caviderm Page 68	Unit Process Assemblies Inc.	D	Micro-resistance	Metal
9	Chemigage meter Fig. 29, page 29	Elcometer Instruments Ltd.	A	Magnetic attraction	Non-ferro-magnetic
10	Coating meter Type 2.099	Institut Dr. Förster	C	Magnetic induction	Non-ferro-magnetic
11	Deposit thickness monitor DTM-4 Page 68	Sloan Instruments Corp.	D	Change of frequency of quartz oscillator	Vapour de-posited layer
12	Dermitron Fig. 53, page 44	Unit Process Assemblies Inc.	C	Eddy current	‡Metal or non metal
13	Diameter coating thickness meter Page 41	Dipl.-Ing. Heinrich List		Magnetic induction	Non-ferro-magnetic
14	Eddy gauge model 133 Fig. 60, page 47	Elcometer Instruments Ltd.	C	Eddy current	Non-conducting
15	Elcometer pull-off gauge Fig. 31, page 30	Elcometer Instruments Ltd.	A	Magnetic attraction	Non-ferro-magnetic

Table 6. Non-destructive Thickness Measuring Instruments (contd.)

Basis material	*Dimensions, Weight, Power supply*	*Instrument characteristics*†
Metal or non-metal		Measuring area 6mm^2-1·3cm^2 Single radiator point Digital presentation
Ferromagnetic	Pencil-shaped	Range 0-400 μm Smallest indication 5 μm Accuracy ± 15%; ± 2·5 μm minimum Single point application
Magnetic or non-magnetic	13 × 7 × 6 in	Measuring area 2·485 cm^2 Range: 25-175 μm (0·5-4 oz/ft^2 zinc on steel) Probe head: centrally mounted detector and two 1 m curie Sr 90 sources
Non-metal	6 kg Mains supply	Minimum hole size 0·25 mm Ranges 0-300 micro-ohms 0-1200 micro-ohms Presentation: meter or digital
Ferromagnetic	2¾″ × ⅝″ (7 × 1·6 cm)	Go-no-go gauge range 0-⅛″ Accuracy 10-15%
Ferromagnetic	280 × 200 × 115 mm^3 4·5 kg Battery supply (life 200 h)	3 ranges, 0-8·5, 4·5-12·5, 10-18·5 mm Accuracy ± 5% of scale value Single pole probe
	Mains supply	Range about 0-3 μm according to material Accuracy ± 2% 5 frequency ranges
Metal or non-metal	430 × 230 × 280 mm^3 12 kg Mains supply	Measuring area 8 mm^2. Internal bores down to 12·5 mm Range 0-2540 μm (according to material) Accuracy ± 5-10% according to material 4 measuring frequencies between 0·1 and 6MHz
Ferromagnetic, non-ferromagnetic or non-metal	Weight 20 kg	Range 0-2 mm according to material Accuracy ± 5%
Non-ferrous	305 × 230 × 140 mm weight 5-3 kg Power supply 115-230 volts, 50-60 Hz Nickel-cadmium battery	3 ranges 0-25 micron (0-0·001 inch) 0-75 micron (0-0·003 inch) 0-250 micron (0-0.10 inch) Accuracy ± 10%
Ferromagnetic	Pencil shape	3 ranges 0-600 micron 0-25 thou 0-10 arbitary Accuracy of ± 15%

[*continued over*

	Name of Instrument	Manufacturer*	Price Category**	Principle of Measurement	Coating
16	Elcometer Thickness Gauge Model 101 Fig. 36, page 33	Elcometer Instruments Ltd.	B	Magnetic induction	Non-ferro-magnetic
17	Elcotector Mk III Fig. 59, page 47	Elcometer Instruments Ltd.	C	Eddy current	‡Metal or non-metal
18	Elektrotest FP Fig. 49, page 40	Elektro-Physik	C	Magnetic induction	Non-ferro-magnetic
19	Elektrotest Standard Model Fig. 47, page 39	Elektro-Physik	C	Magnetic induction	Non-ferro-magnetic
20	Elektrotest U Fig. 48, page 40	Elektro-Physik	C	Magnetic induction	Non-conducting
21	Elmicron FE Fig. 45, page 38	EFCO Ltd., Sel-Rex	C	Magnetic induction	Non-ferro-magnetic
22	Elmicron NF Fig. 62, page 49	EFCO Ltd., Sel-Rex	C	Eddy current	Non-conducting
23	FH 46 Measurement and control system Fig. 75, page 59	Frieseke & Hoepfner GmbH	D	Beta backscatter, Beta trans-mission	¶Metal or non-metal
24	Filmeter NRL Model 1 Page 47	American Instrument Co. Inc.		Eddy current	Non-conducting
25	Film thickness monitor FTM2 Page 68	Edwards Instruments	C	Quartz oscillator	Vacuum evaporated
26	Fluoroscopy gauge Page 66	EKCO Instruments Ltd.	D	Backscatter	¶Metal or non-metal

Table 6. Non-destructive Thickness Measuring Instruments (contd.)

Basis material	*Dimensions, Weight, Power supply*	*Instrument characteristics*†
Ferromagnetic	80 × 50 × 25 mm^3 185 g 3 different models A. Standard B. For round materials C. For soft materials	Round materials, convex from r = 3·2 mm, concave from r = 2·5 mm Ranges 0-50, 0-80, 0-100, 0-200, 0-250, 0-600, 400-1000 μm Accuracy ± 5% of value, down to ± 2·5 μm Two-point probe
Metal or non-metal	152 × 203 × 102 mm^3 3·7 kg Mains or battery Battery life 200 h	3 different measuring frequencies Various probe forms
Ferromagnetic	240 × 150 × 160 mm^3 Battery supply (life 100 h)	Ranges 0-100 μm and 0-2500 μm Accuracy ± 5% of measured value down to ± 2·5 or ± 10 μm Two pole probe
Ferromagnetic	240 × 150 × 160 mm^3 3 kg Mains or battery supply	Measuring area 5 cm^2 Ranges 0-100 or 0-300, 0-3000, 0-10000 μm Accuracy ± 5% down to ± 5 μm Single pole probe
Non-ferrous metal	240 × 150 × 160 mm^3 4 kg Mains supply	Measuring area 5 cm^2 Ranges 0-220 μm, 0-1000 μm Accuracy ± 5% down to ± 20 μm Single pole probe
Ferromagnetic	290 × 210 × 115 mm, 110-250 V, 50-60 Hz, with built-in rechargeable battery	Ranges 0-30 μm, 25-100 μm, 75-300 μm Accuracy better than 5% Measuring area 5 × 1 mm Round material: concave from 5 mm, convex from 0·5 mm
Non-ferrous metal	290 × 210 × 115 mm, 110-250 V, 50-60 Hz, with built-in rechargeable battery	Ranges 0·30 μm, 25-100 μm, 75-300 μm Accuracy better than 5% Measuring area 5 mm diameter
Metal or non-metal	Mains supply	9 ranges: back scatter operation 0-3·10^5 g/cm^2 Accuracy ± 1-10% according to material 7 different radiators Distance between specimen and measuring chamber 20 mm
Non-ferromagnetic metal	5 kg Battery supply	Round material from r = 15 mm Ranges: I 0-25 μm; II 0-126 μm Accuracy, range I ± 5% of full scale, min ± 2·5 μm; range II, ± 3% of full scale
	Mains powered (less than 100 W)	4 frequency ranges, 3 rates of thickness change ranges. Direct indication in Å, range 0-10 kÅ, digital or analogue. Deposition rate and termination in separate modules
Metal or non-metal		Various types of probe Range 1·3-127 μm Accuracy ± 1-10% Source: promethium

[*continued over*

	Name of Instrument	*Manufacturer**	*Price Category***	*Principle of Measurement*	*Coating*
27	HI interference microscope Page 53	Hacker Instruments Inc.	D	Optical interference	Metal or non-metal transparent
28	Inspector thickness Gauge Fig. 27, page 28	Elcometer Instruments Ltd.	B	Magnetic attraction	Non-ferro-magnetic
29	Isometer type 2.082 Fig. 56, page 45	Institut Dr. Förster	C	Eddy current	Non-conducting
29a	Isoscope Page 44	Helmut Fischer GmbH	C	H.F. eddy current	Non-conductive
30	Lectromag Fig. 33, page 31	Lea Manufacturing Co.	B	(Electro) magnetic attraction	Non-ferro-magnetic
31	Leptometer type NET 200 Fig. 57, page 46	Karl Deutsch Erichsen GmbH & Co.	C	Eddy current	Non-conducting
32	Leptoscope SMG Type EP Page 35	Karl Deutsch Erichsen GmbH & Co.	C	Magnetic induction	Non-ferro-magnetic
33	Leptoscope T 500 Fig. 41, page 36	Karl Deutsch Erichsen GmbH & Co.	C	Magnetic induction	Non-ferro-magnetic
34	Leptoscope Universal SMG 8 Fig. 40, page 35	Karl Deutsch, Erichsen GmbH & Co.	C	Magnetic induction	Non-ferro-magnetic
35	Magne-Gage Fig. 28, page 28	American Instrument Co. Inc.	C	Magnetic attraction	a) Non-ferro-magnetic b) Nickel c) Nickel

Table 6. Non-destructive Thickness Measuring Instruments (contd.)

Basis material	*Dimensions, Weight, Power supply*	*Instrument characteristics*†
Metal or non-metal non-transparent		Magnification 100× and 200× Measuring range 0·03-2 micron 35 mm camera attachment
Ferromagnetic	50 × 30 × 230 mm^3 231 g	Range 0-500 μm
Non-ferro-magnetic metal	280 × 200 × 115 mm^3 4 kg Battery (life 500 h)	Measurement area 4 cm^2 Round material from r = 2·5 mm Ranges 0-100 and 0-300 μm Accuracy ± 1·5% of full scale Tripod probe
Ferrous and non-ferrous	185 × 95 × 50 mm 0·75 kg Dry cell battery	Min. measuring area 3 mm Min. diameter of test piece 6 mm Measurement range 2 μm - 350 μm Accuracy: 3%
Ferromagnetic	Length 100 mm Mains supply	Range 5·1-203 μm Accuracy: above 25 μm ± 10% below 25 μm ± 15%
Metal	100 × 70 × 35 mm^3 250 g Battery supply (life 300 h) 2 different models	Pocket model Ranges 0-100 and 0-200 μm Accuracy ± 10% of measured value 4 Single-pole probes
Ferromagnetic	210 × 210 × 300 mm^3 6·5 kg Mains supply 4 different models Models EP single pole probe Models ZP two pole probe	Ranges: Model EP 100/500 I 0-100 μm; II 0-500 μm Model EP 500/2000 I 0-500 μm; II 0-200 μm Model ZP8 1-8 mm Model ZP20 1·5-20 mm Accuracy ± 5% of measured value
Ferromagnetic	70 × 150 × 90 mm^3 2 kg Battery supply 2 different models	Two ranges: Model a 0-50 and 0-500 μm Model b 0-200 and 0-2000 μm Accuracy ± 5% of measured value Single pole probe
Ferromagnetic	210 × 210 × 300 mm^3 6·5 kg Mains supply	Ranges 0-100, 0-500, 0-2000 & 0-8000 μm Accuracy ± 5% of measured value Single probe for ranges I, II, III Two pole probe for range IV
Ferromagnetic Steel Non-ferro-magnetic	220 × 220 × 120 mm^3 453 g Four different magnets: single point application	Measuring area 5 cm^2 Round material (concave and convex) from R = 6 mm Ranges: a) 0-50, 37-200, 180-650, 600-2000 μm b) 0-20, 12-50 μm c) 0-25 μm Accuracy ± 1%

[*continued over*

	Name of Instrument	*Manufacturer**	*Price Category***	*Principle of Measurement*	*Coating*
36	Magnus-Junior Fig. 35, page 33	A. Bergner & Co.	B	Magnetic induction	Non-ferro-magnetic
37	MA S/P magnetic thickness gauge Fig. 32, page 30	Laboratorium Prof. Dr. Berthold	B	Magnetic attraction	Non-ferro-magnetic
38	Micro-Derm Fig. 79, page 62	Unit Process Assemblies Inc.	D	Beta backscatter	¶Metal or non-metal
39	Microscope (double beam interference) Fig. 70, page 54	Ernst Leitz GmbH. Bergmann, K.-G.	D	Optical interference	Transparent (anodic oxide coatings) Steps in any material
40	Microscope (double beam interference) Fig. 69, page 53	Carl Zeiss	D	Optical interference	Transparent (anodic oxide coatings) Steps in any material
41	Microscope (light section) Page 51	Exaphot-Optik GmbH	C	Light section	Transparent
42	Microscope (light section) Fig. 67, page 52	Carl Zeiss	C	Light section	Transparent
43	Mikrotest Fig. 26, page 28	Elektro-Physik	Models G & F—B Electro-plating model—C	Magnetic attraction	Non-ferro-magnetic (nickel only relative)
44	Mini-Leptoscope cf. page 35	Karl Deutsch Erichsen GmbH & Co.	C	Magnetic induction	Non-ferro-magnetic
45	Minitector Fig. 43, page 37	Elcometer Instruments Ltd.	C	Magnetic induction	Non-ferro-magnetic

Table 6. Non-destructive Thickness Measuring Instruments (contd.)

Basis material	*Dimensions, Weight, Power supply*	*Instrument characteristics*†
Ferromagnetic	90 × 80 × 50 mm^3 2 different models	Ranges: Standard model 100-700 μm Electroplate model 0-100 μm Accuracy ± 10% Two point application
Ferromagnetic	Pencil form	Ranges 0-15, 10-40 μm, 30-100, 100-500 or 400-800 μm Accuracy ± 10%
Metal or non-metal	430 × 380 × 230 mm^3 32 kg Mains supply Measuring table model, circuit board surfaces probe and through-plated hole probe	Measuring area 0·06 mm^2 Round material from r = 0·15 mm Range 0-300 μm according to source and material 6 different radiator points Presentation: analogue or digital direct reading
Non-trans-parent	250 × 270 × 450 mm^3	Adjustable magnifications ×100, ×200, ×500
Non-trans-parent	270 × 180 × 190 mm^3 21·5 kg Mains supply	Range 0·03-2 μm Adjustable magnifications ×80, ×200, and ×480 Accuracy ± 5%
Non-trans-parent	27 kg	Range 3-100 μm Adjustable magnifications ×60, ×120, ×260, and ×520
Non-trans-parent	18·5 kg	Range 2·5-400 μm Adjustable magnifications ×200, ×400 Accuracy ± 0·5 μm
Ferromagnetic	215 × 28 × 75 mm^3 115 g 4 different models	Measuring area 1 cm^2 Ranges: Model G 0-50, F 0-500, S5 500-5000, S10 2500-10000 μm Accuracy ± 10% with limiting values, model G ± 1 μm, F ± 10 μm, S5 ± 100 μm, S10 ± 200 μm Single point application
Ferromagnetic	100 × 70 × 35 mm^3 450 g Battery	Range 0-200 μm Accuracy ± 5% of measured value Two pole probe
Ferromagnetic	120 × 95 × 43 mm case 65 × 19 mm diameter probe Power supply PP3 batteries	Single pole probe 2 models ranges 0-125, 0-600 μm or 0-600, 250-1500 μm

[*continued over*

	Name of Instrument	Manufacturer*	Price Category**	Principle of Measurement	Coating
47	Monimeter type 2.094 Fig. 42, page 37	Institut Dr. Förster	C	Magnetic induction	Non-ferro-magnetic
48	Nachet 300 page 53	Hacker Instruments Inc	D	Optical interference and differential interferometry	Metallic and non-metallic and transparent
49	Nickelscope (Permascope type EN) Fig. 55, page 45	Helmut Fischer GmbH	C	Modified electro-magnetic induction	Nickel (not electroless)
50	Norelco tin surface density meter Page 67	Philips Electronic Instruments		X-rays	Tin
51	Norelco zinc/aluminium surface density meter Fig. 85, page 67	Philips Electronic Instruments	D	X-rays	Aluminium or zinc
52	Optical coating thickness meter	Leybold GmbH	C	Optical interference (intensity measurement)	Vapour deposited layer[45–47]
53	Optical film monitor page 56	Edwards Instruments	C	Photoelectric	Vacuum deposits[45–47]
54	Permascope type EC Fig. 54, page 44	Helmut Fischer GmbH	C	H.F. eddy current	Non-conducting; electroless nickel; metal foil
55	Permascope type ES Fig. 44, page 38	Helmut Fischer GmbH	C	Electro-magnetic induction	Non-ferro-magnetic or non-metal; electroless nickel

Table 6. Non-destructive Thickness Measuring Instruments (contd.)

Basis material	*Dimensions, Weight, Power supply*	*Instrument characteristics*†
Ferromagnetic	280 × 200 × 115 mm^3 3 kg Battery (life 200 h)	Round material from r = 1 mm Ranges 0-15, 10-70, 60-260, 250-3000 μm Accuracy ± 3% of full scale Single pole probe
Metallic and non-metallic and transparent	12″ × 9″ × 16″	Range 0·03-2 μm Magnification objectives 13×, 19×, 50× and 70× eye piece 10× 33 mm camera fitment
Ferro-magnetic; Non-ferrous	225 × 114 × 223 mm Models: EN 3·5 kg Models: EN-NB 3·9 kg Mains and battery models	Min. measuring area 4·5 × 1·5 mm Min. diameter of test piece 2 mm Measurement range 1 μm to 125 μm plus extension up to 0·5 mm Accuracy: Dependent upon calibration standards and inherent nickel characteristics
Steel sheet	1778 × 1143 × 610 mm^3 2357 kg Mains supply	Range (single sided) 1·1-6 g/m^2 Accuracy ± 0·1 g/m^2 Measurements can be made on one or both sides Measurement and control system
Steel sheet	1778 × 1143 × 610 mm^3 3000 kg Mains supply	3 ranges for Zn: 15-122, 91-246, 213-462 g/m^2 Range for Al: 76 - 152 g/m^2 Accuracy ± 2% of measured value
Non-transparent	Mains supply	Measuring area 80 mm^2 Range 0·015-10 μm according to material
		Range 4000 Å-25,000 Å with accessories Recorder or process terminator can be attached
Non-ferrous	225 × 114 × 223 Models: EC3 3·6 kg Models: EC-NB3 3·8 kg Mains and battery (NB) models	Min. measuring area 3 mm diameter Min. diameter of test piece 6 mm Measurement range 0·5 μm to 150 μm plus extension up to 70 mm Accuracy: ± 1·5%
Ferromagnetic	225 × 114 × 223 mm Models: ES4 3·8 kg Models: ES4-NB 4·3 kg Mains and battery (NB) models	Min. measuring area 4·7 × 0·8 mm Min. diameter of test piece 0·8 mm Measurement range 0·5 μm - 10 mm plus extension Accuracy: 3%

[*continued over*

	Name of Instrument	Manufacturer*	Price Category**	Principle of Measurement	Coating
56	PW 4390 coating thickness meter Page 67	Philips Industrie Elektronik GmbH	D	Beta backscatter	‡Metal, non-metal
57	Quartz oscillator coating thickness monitor type QC 1 Page 68	Klaus Schaefer GmbH	C	Change of frequency	Vapour deposited (45–47)
58	Quartz oscillator coating thickness monitor type QSG 101 Page 68	Balzers A.-G.		Change of frequency	Vapour deposited (45–47)
59	Reluctance-type thickness gage Page 41	General Electric Co.		Magnetic induction	Non-ferromagnetic
60	SM1 (SM2 and SM3) coating thickness meter Page 41	Dipl.-Ing Heinrich List		Magnetic induction	Non-ferromagnetic
61	Ultrasonoscope film thickness meter Fig. 61, page 48	Ultrasonoscope Co. Ltd	C	Eddy current	‡Non-conducting

---

45. Hacman, D. Optical measurements on metallic vapour-deposited layers in the thickness range up to the limits of transparency. *Balzers Hochvakuum-Fachbericht,* 1965, No. 4.
46. Pulker, H. K. Investigation of continuous thickness measurement of thin vapour-deposited layers with a quartz oscillator measuring equipment. *Z. angew. Phys.,* 1966, **20**, No. 6, 537-40.
47. Pulker, H. and Ritter, E. Short review of methods for determination of thickness of thin layers. *Vakuum-Technik,* 1965, No. 4, 91-97.

Table 6. Non-destructive Thickness Measuring Instruments (contd.)

Basis material	*Dimensions, Weight, Power supply*	*Instrument characteristics*†
Metal or non-metal	Mains supply	Modular system
	Mains	Range about 0-3 μm according to material Accuracy ± 1.5% 4 frequency ranges
	Mains	Range about 0-3 μm according to material Accuracy ± 1·5% 6 frequency ranges
Ferromagnetic	190 × 152 × 152 mm² 3·4 kg Mains	Measuring area 20 cm² Round material: convex from r = 37·5 mm: concave from r = 75 mm 2 ranges selected between 0 and 7 600 μm Accuracy ± 5-10% of calibrated value
Ferromagnetic	275 g 1·5 V battery. Current consumption less than 10 mA	Ranges: 0-2 mm, particularly 20-500 μm, 0-100 μm and 0-12 mm
Non-ferrous	230 × 245 × 460 mm³ 9 kg Mains supply 2 different models	Measuring area 28 mm² Round material from r = 3 mm Range: model ASF/1 for anodic oxide coatings 0-50 μm; model PSF/1 for paint coatings 0-400 μm Accuracy: ASF/1 ± 0·25 μm PSF/1 ± 2·5 μm

Table 7. Standard Specifications

SOME countries, such as Denmark and Austria, have no special standards of their own concerning the measurement of coating thickness.

In any country the Standard Specifications, both National and Foreign, are obtainable from the National Standards Organisation whose address is given in the list for each country.

In this Table, after ISO, the countries are arranged in alphabetical order.

ISO (International Standards Organisation)

Address :
1, rue de Varembé,
1211 - Geneva 20 - Switzerland.

ISO Recommendation	R 1463	Measurement of metal and oxide coating thicknesses by microscopical examination of cross-sections.
Draft ISO Recommendation	2177	Measurement of coating thickness-coulometric method by anodic dissolution.
Draft ISO Recommendation	2178	Measurement of coating thickness—magnetic method—(non-magnetic coating on magnetic basis metal).

CZECHOSLOVAKIA

Address :
Urad pro notmalisaci a mereni,
Prague.

CSN 038 150	Thickness measurement of anodic oxide coatings on aluminium.
CSN 038 156	Coating thickness measurements of electrolytically deposited metallic coating by destructive methods.
CSN 038 157	Coating thickness determinations by non-destructive methods.
CSN 039 046	Chemical coating thickness measurements of tin coatings.

FRANCE

Address :
Association Francaise de Normalisation,
23 rue Notre Dame des Victoires, Paris 2^e^.

NF A 36-321 (1966)	Continuous dip coated coatings on sheets: gravimetric method.
NF A 81-131 (1962)	Dip coatings on steel wires: gas volumetric method.
NF A 91-101 (1964)	Electrolytic nickel and chromium coatings: chemical stripping, sectioning, drop, coulometric, magnetic attraction and parallel flux eddy current methods.
NF A 91-102 (1947)	Electrolytic zinc and cadmium coatings: drop and chemical stripping methods.

NF A 91-105 (1967)	Chemical nickel coatings: sectioning and coulometric methods.
NF A 91-121 (1958)	Dip-coated zinc coatings: gravimetric method.
NF A 91-201 (1966)	Spray metallisation (zinc, aluminium, lead): sectioning, magnetic attraction and parallel flux methods.
NF A 91-301 (1965)	Electrolytic hard chromium coatings: direct measurement before and after coating, magnetic attraction and parallel flux methods.
NF A 91-402 (1965)	Anodic oxide coatings on aluminium: sectioning method.
NF A 91-403 (1966)	Anodic oxide coatings on aluminium: non-destructive microscopic determination.
NF A 91-404 (1965)	Anodic oxide coatings on aluminium: eddy current method.
NF A 91-405 (1966)	Anodic oxide coatings on aluminium: measurement of breakdown voltage.
NF A 91-406 (1966)	Anodic oxide coatings on aluminium: gravimetric thickness measurement.

GERMAN DEMOCRATIC REPUBLIC

Address :
Amt für Standardisierung der DDR,
108 Berlin, Mohrenstr. 37a.

TGL 49-96 901 (1958) TGL 0-50 950 (1962)	Microscope determination of the thickness of electrodeposits.
TGL 49-96 902 (1958) TGL 18 780	Determination of coating thickness by the jet method.
TGL 49-96 903 (1958) TGL 0-50 953 (1962)	Determination of the thickness of thin chromium deposits by the spot test.
TGL 49-96 904 (1958)	Determination of coating thickness by the light section method.
TGL 49-96 905 (1959)	Microscope determination of the thickness of inorganic protective coatings.
TGL107-06 101 (1960)	Testing of paint films; determination of thickness with the clock gauge and micrometer screw gauge.

Table 7. Specifications (continued)

GERMAN FEDERAL REPUBLIC

Address :
Deutsche Normenausschuss,
1 Berlin 30, Burggrafenstr. 4-7.

DIN 50 932 (1971)	Testing of metallic coatings; measurement of the thickness of zinc coatings on steel by local anodic dissolution, coulometric method.
DIN 50 933 (1971)	Draft. Testing of metallic coatings; determination of the thickness of coatings on steel with a dial indicator.
DIN 50 943 (1961)	Draft. Testing of inorganic non-metallic coatings on aluminium and aluminium alloys; microscope determination of coating thickness.
DIN 50 944 (1968)	Testing of inorganic non-metallic coatings on pure aluminium and aluminium alloys; non-destructive measurement of the thickness of transparent oxide layers by the differential method with the microscope.
DIN 50 948 (1967)	Draft. Testing of inorganic non-metallic coatings on aluminium and aluminium alloys; non-destructive measuring of the coating thickness of transparent aluminium oxide coatings by the light section method.
DIN 50 945 (1968)	Testing of inorganic non-metallic coatings on pure aluminium and aluminium alloys; non-destructive measurement of the thickness of transparent oxide layers by the differential method with the microscope.
DIN 50 950 (1968)	Testing of electroplated coatings; microscope measurement of coating thickness.
DIN 50 951 (1971)	Draft. Measurement of the thickness of Electroplated metallic coatings by the radiation method.
DIN 50 952 (1963)	Testing of metallic coatings; determination of surface density of zinc coatings on steel by chemical stripping; gravimetric method.
DIN 50 953 (1967)	Electroplated deposits; determination of the coating thickness of thin chromium layers by the spot test.
DIN 50 954 (1965)	Testing of metallic coatings; determination of the mean surface density of tin coatings on steel by chemical stripping.
DIN 50 955 (1971)	Testing of metallic coatings; measurement of the thickness of electroplated metallic coatings; coulometric method.
DIN 51 213 (1965)	Pre-standard. Testing of metallic coatings on wires.

TABLE 7. SPECIFICATIONS

GREAT BRITAIN

Address :
British Standards Institution,
British Standards House, 2 Park St., London W1.

B.S. 443: 1969	Galvanized coatings on wires. Methods: gravimetric, gas volumetric, Preece.
B.S. 729: 1961 (with draft 31596 and ammendment AMD 395: 1969)	Zinc coatings on iron and steel articles. Part 1. Hot dip galvanized. Methods: gravimetric, Preece. Part 2. Sherardised coatings. Method: Preece.
B.S. 1224: 1970	Electroplated coatings of nickel and chromium. Methods: sectioning, coulometric.
B.S. 1615: 1961	Anodic oxidation coatings on aluminium. Methods: sectioning, gravimetric, breakdown voltage, coulometric.
B.S. 1706: 1960	Electroplated coatings of cadmium and zinc on iron and steel. Methods: jet, gravimetric.
B.S. 1872: 1964	Electroplated coatings of tin. Methods: sectioning, gravimetric.
B.S. 2816: 1957	Electroplated coatings of silver for engineering purposes. Methods: sectioning, jet.
B.S. 2989: 1967	Hot-dip galvanized plain steel sheet and coil. Method: gravimetric.
B.S. 3083: 1959	Hot-dip galvanized corrugated steel sheets for general purposes. Methods: gravimetric, weighing of a blind test sheet moving through the bath.
B.S. 3382: 1961	Electroplated coatings on threaded components. Part 1. Cadmium on steel parts. Part 2. Zinc on steel parts. Method: gravimetric.
B.S. 3597: 1963	Electroplated coatings of 65/35 tin-nickel alloy. Methods: sectioning, coulometric.

HUNGARY

Address :
Szabványbolt,
Budapest V, Szt. istván ter 4.

MSz 6575/3-61	Testing of electrolytic deposits: coating thickness measurement. The Standard details jet and drop methods, determination by weight difference, and the sectioning method.
MSz 7741/1-61	Testing of electrolytic oxide layers: coating thickness measurement. The Standard details determination by stripping the layer and by the microscope method.
MSz 9648-59	Thickness measurement of varnish coatings.

Table 7. Specifications (continued)

ITALY

Address :
Ente Nazionale Italiano di Unificazione,
Piazza Diaz 2, Milano.

UNI 3396 (1953)	Thickness determination of anodic oxide coatings by weighing.
UNI 4115 (1959)	Testing of thickness of anodic oxide coatings with the aid of break-down voltage.
UNI 4195 (1959)	Rules for determination of thickness of metallic coatings by magnetic methods.
UNI 4237 (1959)	Sectioning methods.
UNI 4238 (1959)	Jet methods.
UNI 4526 (1960)	Drop method.
UNI 4528 (1960)	Weight determination of phosphate coatings.
UNI 5084 (1962)	Weight determinations of electrolytically deposited coatings of silver and copper and their alloys.
UNI 5345 (1964)	Chemical determination of mean coating thickness of tin, copper and their alloys.
UNI 5347 (1964)	Thickness determination of non-conducting coatings on non-ferrous metals by electromagnetic methods.

NETHERLANDS

Address :
Nederlands Normalisatie Instituut,
Polakweg 5, Rijswijk (Z.H.)

NEN 2169 (1959)	Coating layers on metals. Microscope determination of local coating thickness.
NEN 5252 (1963)	Determination of coating thickness by the jet method.
NEN 5255 (1959)	Anodic oxidation of aluminium: coating thickness.

NORWAY

Norges Standardiserings-Forbund,
Haakon VII's gt.2, Oslo 1.

NS 1181 (1964)	Metal coating thickness determination with the microscope.
NS 1183 (1964)	Metal coating thickness determination by dissolving off the layer.

NS 1185 (1964)	Metal coating thickness determination of thin chromium layers by the spot test procedure.

POLAND

Address :
Polski Komitet Normalizacyjny,
Warszawa, ul. Wiejska 20.

PN-58/H-04 614	Determination of coating thickness of thin chromium deposits by the spot test procedure.
PN-58/H-04 616	Microscope determination of coating thickness.
PN-59/C-81 515	Electromagnetic thickness measurement of varnish layers.
PN-61/H-04 605	Measurement of thickness of electrolytically deposited coatings: chemical and electrolytic stripping.

RUMANIA

Address :
Officiul de stat pentru Standarde,
Str. Edgar Quinet 6, Raion 30 Dec., Bucharest.

STAS 2951-51	Determination of thickness of tin coatings.
STAS 6853-63	Determination of coating thickness by the jet method.
STAS 6854-63	Determination of coating thickness by the drop test.
STAS 7043-64	Electrochemical oxidation of aluminium and aluminium alloys: technical regulations: measurement of breakdown voltage.
STAS 7221-65	Zinc coatings on steel and cast iron by the hot dip process. (Gives general regulations, including measurement of coating thickness.)

SWEDEN

Address :
Swedish Corrosion Institute,
Grevturegatan 14, Box 5073, Stockholm 5.

K 520	General on coating thickness and uniformity testing.
K 5212	Mechanical testing and measurement.
K 522	Magnetic testing.

[*continued over*

Sweden (continued) **Table 7. Specifications (continued)**

K 5231 Chemical testing: absolute methods, stripping and analysis of zinc coatings.

K 5232 Chemical testing: absolute methods, stripping and analysis of cadmium coatings.

K 5233 Chemical testing: absolute methods, stripping and analysis of aluminium coatings.

K 5235 Chemical testing: absolute methods, stripping and analysis of lead coatings.

K 5243 Chemical testing: absolute methods, including anodic oxide coatings on aluminium.

K 5250 General on chemical testing: relative methods, jet method, local determination.

K 5254 Chemical testing: relative methods, jet method, local determination of tin coatings.

K 5256 Chemical testing: relative methods, jet method, local determination of nickel and copper coatings.

K 5261 Chemical testing: relative methods, immersion method, local determination of tin coatings.

K 5267 Chemical testing: relative methods, immersion method, local determination of chromium coatings.

K 5271 Inductive methods for measurement of anodic oxide coatings on aluminium.

SWITZERLAND

Address :
VSM-Normalienbureau,
Zürich.

VSM 31a/3 Corrosion protection by metallic coatings.

Draft Determination of coating thickness.
A. Separation methods: sectioning measurements with the microscope
B. Other methods.
1.1 Oblique section
1.2 Dissolution of the basis metal
2.1 Clock gauge method
2.2 Drop, jet and spot test methods
2.3 Coulometric method
3.1 Stripping of the coating and determination of weight loss
3.2 Stripping of the coating and analytical determination of coating metal in the solution
3.3 Calorimetric method
4.1 Magnetic methods
4.2 Electromagnetic methods
4.3 Eddy current method
4.4 Thermoelectric method
4.5 Beta-ray methods.

VSM 37 251 Anodic oxidation of aluminium and aluminium alloys. Coating thickness

Microscope measurement of thickness on sectioned samples
Measurement with eddy current instruments
Measurement with breakdown voltage instruments
Measurement with resistance measuring instruments
Measurement by the light section method.

UNITED STATES OF AMERICA

Address :
American Society for Testing and Materials,
1916 Race St., Philadelphia, Penna. 19103; and
National Bureau of Standards, Washington 25, DC.

ASTM A 90	Drop test for zinc coatings on iron
ASTM A 219-58	Test for local thickness of electrodeposited coatings (sectioning, drop and spot tests and magnetic methods).
ASTM A 309–54	Determination of the weight and structure of coatings on tinware by the spot test.
ASTM A 464	Recommendations for measuring zinc coatings on iron and steel with the help of magnetically operated instruments.
ASTM B 137–45 (1965)	Test for weight of coating on anodically coated aluminium.
ASTM B 244-62 T	Tentative method of measuring thickness of anodic coatings on aluminium with eddy current instruments.
ASTM D 1005–51 (1966)	Method of measurement for paints, varnish, lacquers, and similar finishes.
ASTM D 1185–53 (1961)	Thickness measurement of dry, non-magnetic layers of paint, varnish, lacquer, and similar finishes on magnetic base materials.
ASTM D 1212–54 (1965)	Thickness determination of wet layers of paint, varnish, lacquers and similar finishes.
ASTM D 1400-58	Thickness determination of dry, non-magnetic layers of paint, varnish, lacquer and similar finishes on non-magnetic base materials.
ASTM E 216-63 T	Tentative recommended practice for measuring coating thickness by magnetic or electromagnetic methods.

USSR

Address :
Library of the USSR,
W. I. Lenin, Moscow Central.

GOST 3003-58	Electroplated and decorative protective coatings. Methods of checking the thickness of zinc, cadmium, copper, nickel and multi-layer coatings.
GOST 3263-46	Method for chemical thickness checking of tin coatings.

Table 8. Names and Addresses of instrument manufacturers and their associates and local agents

1. M. L. Alkan Ltd., Ruislip, Middlesex, England.
2. American Instrument Company, Inc., 8030 Georgia Ave., Silver Spring, Maryland 20910, U.S.A.; England — Elcometer Instruments Ltd.; Germany — Erichsen GmbH; Italy — Ingg. S. & Agostino Belotti.
3. Automation Industries Inc., Shelter Rock Road, Danbury, Connecticut 06810, U.S.A.
4. Balzers A.-G., F1-9496 Balzers, Liechtenstein.
5. Bell Telephone Laboratories Inc., Holmdel, N.J. 07733, U.S.A.
6. Ingg. S. & Agostino Belotti, Milano (718), Piazza Trento 8, Italy.
7. Ing. S. & Dr. Guido Belotti, Milano, Piazza Trento 8, Italy.
8. Franz Bergmann KG, 1 Berlin 37, Berliner Strasse 25, Germany.
9. A. Bergner & Co., Bern, Wankdorffeldstrasse 92/92a, Switzerland.
10. Laboratorium Prof. Dr. Berthold, 7547 Wildbad, Calmbacher Str. 22, Postfach 160, W. Germany.
11. BFI-Elektronik GmbH., 6 Frankfurt/Main, Hainer Weg 26-28, Germany.
12. Em. & P. Bodson (P & F), Appareils de Laboratories, 6 Quai Saint-Léonard, Liège, Belgium.
13. British Drug Houses Ltd., Poole, Dorset, England.
14. Bundesanstalt für Materialprüfung, 1 Berlin 45, Unter den Eichen 87, Germany.
15. Burndept Ltd., Erith, Kent, England.
16. Candorchemie GmbH, 463 Bochum, Postfach 144, Germany.
17. Carobronze Ltd., School Road, Belmont Road, London W.4, England.
18. Detakta, 2 Hamburg 39, Alsterdorfer Strasse 266, Germany.
19. Karl Deutsch, 5600 Wuppertal 1, Postfach 3115, W. Germany; England — Vitosonics Ltd.; Germany — Erichsen GmbH & Co.
20. Dressler Elektronik, 1 Berlin 30, Postfach 100, Germany and 2804 Lilienthal/Bremen, Gutenbergstr. 6, Germany.
21. Edwards Instruments Ltd., Manor Royal, Crawley, Sussex, England.
22. EKCO Instruments Ltd., Southend-on-Sea, Essex, England.
23. Elcometer Instruments Ltd., Fairfield Road, Droylsden 38, Manchester; Germany — Erichsen GmbH & Co.
24. Elektro-Physik, Hans Nix & Dr.-Ing. E. Steingroever K.G., 5 Köln-Weidenpech 1, Postfach 351; England — Rubert & Co. Ltd., Acru Works, Demmings Road, Cheadle, Cheshire.

24a. Engelhard Industries (France), 48 Av. Victor-Hugo, 75-Paris 16.

25. Erichsen GmbH & Co., 587 Hemer-Sundwig/Westf., Am Iserbach, Germany; England — Elcometer Instruments Ltd.
26. Evershed & Vignoles Ltd., Acton Lane, Chiswick, London, W.4., England.
27. Exaphot-Optik GmbH, 1 Berlin 15, Uhlandstrasse 158, Germany.
28. Exatest GmbH, 509 Leverkusen, Friedrichstr. 38, W. Germany.

29. M. Falk & Co. Ltd., Reigate, Surrey, England.

30. Helmut Fischer GmbH & Co., 7034 Maichingen, Industrie-Strasse 21, Germany; Australia — Pyrox Ltd.; Belgium — Em. & P. Bodson (P & F); Denmark — B. Zacharaisson; Finland — Havulinna OY; France — Testwell S.A.; India — Motwane (Private) Ltd.; Italy — Ing. S. & Dr. G. Belotti; Japan — Hissan Trading Co. Ltd.; Netherlands — N.V. Electrotechnische Mij.; Norway — Fly & Industrie; Sweden — H. Wallenberg & Co. AB; Switzerland — Langbein-Pfanhauser Werke AG; U.K. — Fischer Instrumentation (G.B.) Ltd.; U.S.A. — Twin City Testing Corp.

31. Fischer Instrumentation (GB) Ltd., Arnhem Road, Industrial Estate, Newbury, Berks., England.

32. Fly & Industrie, Instrumenter A.S., Wm. Thranes Gate 84B-86, Oslo, Norway.

33. Institut Dr. Förster, 7410 Reutlingen (Württ.), Grathwohlstrasse 4, W. Germany; U.S.A. — Automation Industries Inc.

34. Frieseke & Hoepfner GmbH, 852 Erlangen-Bruck, Schliessfach Nr. 72, Germany.

35. Gardner Laboratories Inc., P.O. Box 5728, Bethesda, Md. 20014, U.S.A.; Germany — Erichsen GmbH & Co.

36. General Electric Company, 159 Madison Ave., New York, N.Y. 10016, U.S.A.

37. Hacker Instruments Inc., Box 646, W. Caldwell, N.J. 07006, U.S.A.

38. Havulinna OY, Vuorikatu 16, Helsinki, P.O. Box 10468, Finland.

39. Hissan Trading Co. Ltd., Daini Maruzen Bldg., No. 2, 3 - Chome Edobashi, Nihonbashi, Chuo-Ku, Tokyo, Japan.

40. Imasa-Silvercrown, 660 Ajax Avenue, Trading Estate, Slough, Bucks., England.

41. IWEG GmbH, 4 Düsseldorf, Postfach 4508, Germany.

42. Jenoptik Exaphot GmbH, 1 Berlin 15, Uhlandstrasse 158, Germany.

43. Johnson, Matthey Metals Ltd., 81 Hatton Garden, London EC1P 1AE, England.

44. Keller u. Bohacek K.G., 4000 Düsseldorf-Rath, Postfach 45, Germany.

45. Kocour Company, 4800 S. St. Louis Avenue, Chicago, Illinois 60632, U.S.A.

46. Koehler Instrument Co. Inc., 168-56 Douglas Avenue, Jamaica, New York 11433, U.S.A.

47. H. C. Kröplin GmbH, 649 Schlüchtern, Auf der Röthe, West Germany; England — Carobronze Ltd.

48. Langbein-Pfanhauser Werke A.-G., 1131 Wien, Postfach 36, Austria.

49. Langbein-Pfanhauser Werke A.-G., 4040 Neuss/Rhein, Postfach 317, Germany.

50. Langbein-Pfanhauser Werke A.-G., Schaffhauserstr. 228, Zurich, Switzerland.

51. The LEA Manufacturing Co., 237 East Aurora St., Waterbury, Conn. 06720, U.S.A.; Lea Manufacturing Co. of England Ltd., P.O. Box 1, Buxton, Derbyshire, England.

52. Ernst Leitz GmbH, 633 Wetzlar, Postfach 210/211, Germany; Germany — Franz Bergmann KG.

53. Leybold GmbH., 5 Köln-Bayenthal, Germany.

54. Dipl.-Ing. Heinrich List, 7022 Oberaichen, Viehweg 17-19, W. Germany.

55. Carl Mahr, 73 Esslingen/Neckar, Postfach 147, Germany.

56. Microscopes Nachet, 17 Rue Saint-Severin, Paris V, France; England — W. Watson & Sons Ltd.; U.S.A. — Hacker Instruments Inc.

57. Motwane (Private) Ltd., 127 Mahatma Ghandi Road, Fort, Bombay, India.

58. C. H. F. Mueller Co., 2 Hamburg 1, Alexanderstrasse 1, Germany.

59. Nash and Thompson Ltd., Hook Rise, Tolworth, Surrey, England.

60. Nuclear Enterprises Ltd., Bath Road, Beenham, Reading RG7 5PR, England.

61. N.V. Electrotechnische Mij., Gebr. van Swaay, Stadhouderslaan 16-18, The Hague, Holland.

62. Panax Equipment Ltd., Holmethorpe Industrial Estate, Redhill, Surrey, England; England — Johnson Matthey Metals Ltd.

63. Dr.-Ing. Perthen GmbH, 3 Hannover 1, Postfach 4720, Germany.

64. Philips Electronic Instruments, 750 South Fulton Avenue, Mount Vernon, N.Y. 10550, U.S.A.

65. Philips Industrie Elektronik GmbH, 2 Hamburg 63, Röntgenstrasse 22, Germany.

65a. Picker Industrial, 1020 London Rd., Cleveland, Ohio.

66. Pyrox Ltd., 18-40 Queensberry St., Melbourne, Australia.

67. Theodor Rapp KG, 7612 Haslach/Kinzigtal, Postfach 61, Germany.

68. R.E.C.A.S. s.p.A., Milano, Viale Gran Sasso 23, Italy.

69. Riedel & Co., 48 Bielefeld, Postfach 6920, Germany.

70. Rubert & Co. Ltd., Acru Works, Demmings Road, Cheadle, Cheshire, England.

71. Klaus Schaefer GmbH, 6 Frankfurt-Niederrad, Postfach 129, Germany.

72. Sheen Instruments (Sales) Ltd., 9 Sheendale Road, Richmond, Surrey, England.

73. Sloan Instruments Division, P.O. Box 4608, Santa Barbara, Calif., U.S.A.

74. Günter Stierand KG, 2863 Ritterhude/Bremen, Postfach 2, Germany.

75. Stierand Prüfgeräte GmbH & Co. KG, 5870 Hemer-Sundwig/West., Germany.

76. Ströhlein & Co., 4 Düsseldorf 1, Postfach 7829, Anderstr. 29, West Germany; England — BCM/875, London W.C.1., England.

77. Testwell S.A., 36 bis rue de la Tour d'Auvergne, Paris 9, France.

78. Thorn Bendix Ltd., Applied Electronics Division, Wellington Crescent, New Malden, Surrey, England.

79. Twin City Testing Corp., P.O. Box 248, Tonawanda, N.Y. 14150, U.S.A.

80. Ultrasonoscope Co. (London) Ltd., Sudbourne Road, Brixton Hill, London S.W.2., England; Europe — M. Falk & Co. Ltd.; U.S.A. — Picker Industrial.

81. Unit Process Assemblies Inc., 53-15 37th Ave., Woodside 77, New York 11377, U.S.A.; England — Imasa-Silvercrown; France — Engelhard Industries; Germany — Dressler Elektronik; Italy — R.E.C.A.S. s.p.A.

82. Vitosonics Ltd., 20 Bull Plain, Hertford, Herts., England.

83. H. Wallenberg & Co. AB, Birger Jarisgatan 4, Stockholm, Sweden.

84. W. Watson & Sons Ltd., Barnet, Herts., England.

85. B. Zachariasson, Skt. Jorgens Alle 8, Copenhagen, Denmark.

86. Carl Zeiss, 7082 Oberkochen, Postfach 35/36, W. Germany.

87. Zormco Electronics Corp., P.O. Box 4444, Cleveland, Ohio, U.S.A.

SUBJECT INDEX

Accuderm 38, 96
Addresses 118
Alkan Coulometric Plating Thickness Meter 21, 88
Alpha-Ray Methods 57
Aluminium Coatings
- chemical solution 70
- guide to non-destructive tests 92
- strip line control by X-ray fluorescence 67
- thickness determination specifications 111, 116

Aluminium Foils
- thickness measurement 58

American Instrument Co. Inc.
- Filmeter 47, 100
- Magne-Gage 28, 102

American Specifications 117
Angstrometer 96
Anodic Oxide Coatings
- methods and instruments for measuring 7, 23, 49, 51, 55, 73, 84, 92
- thickness determination specifications 110, 111, 113, 114, 116, 117

Anodic Stripping Methods 8, 17, 23, 27, 82, 112
Arc Section Method 7
ASTM A 219-58 1, 15
ASTM D 1005-51 9
ASTM Specifications 117
Atomat
- beta backscatter gauge 60, 96
- differential beta-transmission gauge 58, 96
- X-ray fluorescence gauge 67, 96

Atomic Number
- difference necessary for measurement by radiation methods 57

Backing Coatings
- in sectioning methods 1

Baier, S. W.
- 'The BNF Plating Gauge' (reference) 17

Balzers, A.-G.
- Quartz oscillator thickness monitor 68, 108

Basis Metal
- chemical dissolution of 12
- coating/basis metal combinations, ease of measurement 26, 42, 57
- For difficulties associated with the measurement of coatings on particular basis materials see Table 3 (p. 84) column 2 and Table 5 (p. 92) column 2.

Bell Telephone Laboratories Laboratory Tester 24, 88
Bendix Method
- anodic stripping of tin 23

Bergmann, F. see Bogenschütz, A. F.
Bergmann, K.-G.
- double beam interference microscope 54, 104

Bergner, A., and Co.
- Magnus Junior 32, 104

Berthold, Laboratorium Prof. Dr.
- MA S/P magnetic thickness gauge 30, 104

Berthold, R.
- 'A handy coating thickness meter of high accuracy' (reference) 31

Beta 750 Plating Thickness Gauge 64, 98
Beta Backscatter Methods 58
Betameter 63, 96
Beta-Ray Methods 57, 116
Betascope 60, 96
Beta Transmission Methods 57
BFI-Elektronik GmbH
- Kocour apparatus 88

Bierwirth, G.
'X-ray operational control of soft nitrided components of C 15 steel' (reference) 66
B.N.F. Coulometric Thickness Meter
See Coulometric Methods
B.N.F. Jet Test see Jet Tests
B.N.F. Thermo-Electric Thickness Meter 50
Boeing Company
originators of the Betascope 60
Bogenschütz, A. F., Bergmann, F. and Jentzsch, J.
'Non-destructive coating thickness measurement by an interference method' (literature reference) 55
Brass Coatings
anodic solution 76
chemical solution 70
guide to destructive tests 84
guide to non-destructive tests 92
Breakdown Voltage Method 23, 111, 113, 114, 115, 116
Brenner, A. see Brodell, F. P.
British Specifications 113
Britton, S. C. see Hohre, W. E.
Brodell, F. P. and Brenner, A.
'Further studies of an electronic thickness gage' (reference) 43
Bronze Coatings
anodic solution 76
guide to destructive tests 84
guide to non-destructive tests 92
jet test solution for 13
specification 114
BS 1223-1959 15
BSA-Tinsley Pencil Gauge 29, 98
Burndept Gauge 65, 98

Cadmium Coatings
anodic solution of 76
calorimetric method for 16
chemical solution of 70
coulometric method, electrolytes for 22
guide to destructive tests for 85
guide to non-destructive tests for 92
jet test solution for 14
radiation method for 65
specifications 110, 113, 116, 117
Calorimetric Method 16, 116
Capacitative Methods 49
Caviderm 68
Chemical Stripping Methods
specifications including 110, 112, 113, 114, 115, 116
stripping basis metal 12
stripping coating 8, 11, 27, 70
Chemigage 29, 98
Chromate Films
cathodic solution 82
chemical solution 71
Chromium Coatings
anodic solution 76
by Kocour apparatus 20
chemical solution 71
coulometric method for 16, 17
coulometric method, electrolyte for 22
guide to destructive tests 85
guide to non-destructive tests 92
jet test solution for 15
spot test for 15, 112
strip line control by X-ray fluorescence 67
specifications 110,111,112,113,115,116
Chromium-Nickel-Copper Coatings
anodic solution 78
Clark, S. G.
jet tests (reference) 13
Clock Gauges 2, 7, 27, 111, 116
Coating-Basis Metal Combinations
ease of measurement 26
ease of measurement by eddy current methods 42
ease of measurement by radiation methods 57
measurement possible by beta backscatter method 65
Coating Meter Type 2.099 (Förster) 98
Coating Thickness Meters S 1566 and S 1567 9, 88
Cobalt
affecting thermoelectric measurement of nickel thickness 50
Cobalt Coatings
guide to destructive tests 85
guide to non-destructive tests 92
jet test solution for 13
Conductivity see Electrical Conductivity

Copper Coatings
  anodic solution 78
  chemical solution 71
  coulometric method, electrolyte for 22
  guide to destructive tests 85
  guide to non-destructive tests 92
  jet test solution for 13
  specifications 114, 116, 117
Copper-Nickel-Chromium Coatings
  anodic solution 78
Copper-Nickel Coatings
  chemical solution 72
Coulometric Methods
  electrolytes for 22
  methods and equipment for 17
  specifications including 110, 112, 113, 116
Coulometric Plating Gauge Mk III (Thorn Bendix) 21, 88
Coulometric Plating Thickness Meter (Alkan) 21, 88
Cross-Sectioning Method see Sectioning Methods
Crystal Oscillation Methods 68
Czechoslovakian Specifications 110

Dektak 10, 88
Deposit Thickness Monitor (Sloan Instruments Corp) 68, 98
Dermitron 43, 98
Destructive Tests
  principles 1
  general specifications 110
  Methods:
    Bendix 23
    breakdown voltage measurement 23
    calorimetric 16
    chemical dissolution of the basis metal 12
    chemical stripping 11
    clock gauge 7
    electrolytic stripping (coulometric)
    immersion and spot 15, 17
    jet and drop 13
    Mesle chord 13
    sectioning 1
    spectrographic 24
    using wet and/or dry film meters 2
  Equipments:
    BNF-Jet 13
    calorimetric 16
    clock gauge 7
    coulometric 17
    Elcometer wet film thickness gauge 4
    film meters or gauges 2
    Gardner Penetrometer 4
    Gardner wet coating thickness gauge 3
    Gratometer 24
    IG-Watch 9
    Kocour 19
    Kröplin Coating Thickness Meter S 1566 9
    Laboratory Tester 24
    Millimess 8
    Paint Inspection Gauge 6
    PEG Thickness Gage 5
    Pfund Film Thickness Meter 4
    Platimeter 8
    Rossmann Dry Film Thickness Gauge type 233 9
    Rossmann Wet and Dry Film Thickness Gauge type 296 9
    Rossmann Wet Film Thickness Gauge 3
    Seddon 19
    Tolerator 8
    Zaba 11
Detakta
  Pencil Thickness Gauge 98
Deutsch, Karl
  Leptometer 46, 102
  Leptoscopes 35, 102
  Mini-Leptoscope 104
Deutsche, V.
  'New developments in the field of magnetic coating thickness measurement' (reference) 32
Dial Gauges 2, 7, 27, 111, 116
Diameter Coating Thickness Meter 41, 98
Diesing, P. and Schneider, H.
  'Two new laboratory methods for testing electrolytic deposits' (reference) 11
Difference in Height Methods 8, 27, 110, 111
Difference in Weight Methods 8, 11, 12, 27, 110, 113, 1114, 116, 117

Differential Beta Transmission Methods 57
DIN 50 932 18
DIN 50 950 1
DIN 50 951 13
DIN 50 953-1968 15
DIN 50 955 18
Drop Methods 15, 110, 113, 114, 115, 116, 117
Dry Film Meters or Gauges 3, 7, 27
Dry Paint Coatings
thickness determination specification 117
Dühmke, M.
'Coating thickness measurement with the help of optical interference' (reference) 54
Dutch Specifications 114

East German Specifications 111
Eberspächer, O.
'Non-destructive X-ray determination of the thickness of plane parallel surface coatings' (reference) 66
Eddy Current Methods 42, 92, 93, 94, 110, 111, 116, 117
Eddy Gauge Model 133 (Elcometer) 47, 98
Edwards Instruments Ltd.
film thickness monitor 68, 100
optical film monitor 56, 106
EFCO Ltd.
Betameter 63, 96
Elmicron FE 38, 100
Elmicron NF 48, 100
EKCO Instruments Ltd.
Fluoroscopy Gauge 66, 100
Elcometer Instruments Ltd.
Chemigage 29, 98
Eddy Gauge Model 133 47, 98
Elcotector 47, 100
Inspector Thickness Gauge 28, 107
Minitector 37, 104
Minitor 46, 104
Paint Inspection Gauge 6, 90
Pull-Off Gauge 29, 98
Thickness Gauge Model 101 (magnetic induction) 32, 100
Wet Film Thickness Gauge 4, 90
Elcotector 47, 100
Electrical Conductivity
difference necessary for measurement by eddy current method 42
Electrical Induction Methods 32, 93, 94
Electrical Resistance-Measuring Instruments 68, 116
Electrolytic Solution Methods (see also Coulometric Methods) 8, 17, 27, 82, 112, 115, 116
Elektro-Physik
Elektrotest 39, 100
Mikrotest 28, 104
Elektrotest 39, 100
Elmicron 38, 48, 100
Elmymeter 88
Elssner, G.
'The non-destructive determination of the thickness of anodic layers with the interference microscope' (reference) 54
Elze, J. and Pantzke, D.
'Metallographic sections of electrolytically deposited metal coatings' (reference) 1
Equipments see Destructive Tests; Non-Destructive Tests; and the individual equipments.
Erichsen GmbH & Co.
Leptometer 46, 102
Leptoscope 35, 36, 40, 102
Mini-Leptoscope 104
Rossmann wet and dry thickness gauges 3, 9, 90
Erichson, A. M.
wet coating thickness gauge 3
Evershed and Vignoles Ltd.
pencil thickness gauge 29, 98
Exaphot-Optik GmbH.
light section microscope 51, 104

Ferromagnetic Coatings
guide to non-destructive tests 92
FH 46 Measurement and Control System 59, 100
Filmeter 47, 100
Film Meters or Gauges 2, 7, 27, 111, 116
Film Thickness Monitor (Edwards Instruments Ltd.) 268, 100

Fischer, H. and Rupp, H.
'Non-destructive coating thickness measurement with the help of magnetic methods' (reference) 35
Fischer, Helmut, GmbH
Betascope 60, 96
Isoscope 44, 102
Nickelscope 45, 106
Permascope 38, 44, 106
Fizeau Arrangement
for multiple beam interferometry 54
Fluoroscopy Gauge 66, 100
Foils
thickness measurement 58
Förster, Institut Dr.
Coating Meter type 2.099 98
Isometer 45, 102
Monimeter 36, 106
Francis, -.
Seddon test apparatus 19
French Specifications 110
Frieseke & Hoepfner GmbH
Measurement and Control System FH 46 59, 100

Galvanised Structural Steel
checking zinc thickness with Burndept non-destructive thickness gauge 65
specifications 110, 113
Galvanised Wire and Strip (see also Zinc Coatings)
specifications 110, 112, 113
thickness of zinc, by Zaba apparatus 11
by potential-time diagrams 18
by X-ray fluorescence 67
Gamma-Ray Methods 57
Gardner Laboratories Inc.
PEG Thickness Gage 5, 90
Penetrometer 4, 90
Wet Coating Thickness Gage 3, 90
Gas Volumetric Test
specification 113
Gehrke, H. see Katz, W.
General Electric Co.
Reluctance-type Thickness Gage 41, 108
German Democratic Republic Specns. 111
German Federal Republic Specns. 112
Germanium
measurement of oxide layer on 55
Gold Coatings
anodic solution 78
by chemical solution of basis metal 12
by radiation methods 65
chemical solution 72
guide to destructive tests 85
guide to non-destructive tests 92
Gratometer 24, 88
Great Britain
specifications 113
Gühring, H.
'The application of non-destructive coating thickness measuring methods' (reference) 38

Hacker Instruments Inc.
interference microscope 53, 102, 106
Hacman, D.
'Optical measurements on metallic vapour-deposited layers in the thickness range up to the limits of transparency' (reference) 108
Hadert, H.
'Thickness measuring instruments' (reference) 2
Heat of Reaction Method 16
Herlitze, K.
'Thickness and density measurement with radioactive rays' (reference) 59
Hess, B.
'Coating thickness measurement with X-rays' (reference) 66
HI Interference Microscope 53, 102, 106
Hoare, W. E. and Britton, S. C.
'Tin plate testing' (reference) 18
Hot Dipped Zinc Coatings see Galvanised Structural Steel; Galvanised Wire and Strip; Zinc Coatings
Hungarian Specifications 113
Hygroscopic Materials
difficulty of applying capacitative methods to 49

IG-Watch 9
Illig, W.
'The measurement of anodic coatings on aluminium' (reference) 5

Immersion Tests 15
Inclined Section Method 2
Induction Methods see Magnetic Induction Methods
Inspector Thickness Gauge 28, 102
Institut Dr. Forster
  Coating Meter type 2.099 98
  Isometer 45, 102
  Monimeter 36, 106
Instruments see Destructive Tests; Non-Destructive Tests; and the individual instruments.
Insulating Coatings
  guide to non-destructive tests 92, 94
  paint see Paint Coatings
  thin see Anodic Oxide Coatings
Interference (Optical) Methods 53
Isometer 45, 102
Isoscope 44, 102
Italian Specifications 114

Jentzsch, J. see Bogenschütz, A. F.
Jet Tests 13, 88, 111, 112, 113, 114, 115, 116
Johnson Matthey Metals Ltd.
  Beta 750 Plating Thickness Gauge 64, 98

Kalpers, H.
  'Determination of the coating thickness of tin deposits' (reference) 19
Katz, W. and Gehrke, H.
  electrolytic cell for potential-time diagrams 18
Keller and Bohacek
  Zaba apparatus 11, 90
Kocour Apparatus 19, 88
Koehler Instrument Co.
  Pfund film thickness meter 4, 90
Krijl, G. and Melse, J. L.
  'Calorimetric determination of metal coating thicknesses on small objects' (reference) 16
Kröplin Coating Thickness Meter S 1566 9, 88
Kutzelnigg, A.
  'The testing of metallic coatings' (reference) 70

Laboratorium Prof. Dr. Berthold
  MA S/P Magnetic Thickness Gauge 31, 104
Laboratory Tester (Bell Telephone Laboratories) 24, 88
Lacquer Coatings (Transparent) (see also Organic Coatings)
  guide to non-destructive tests 92, 94
  light section method 52
Lakshmanan, A. S. see Mathur, P. B.
Langbein-Pfanhauser Werke, AG., Neuss
  jet test apparatus 13, 88
Langbein-Pfanhauser Werke AG., Vienna
  Gratometer 24, 88
Langel, H.
  coating control with radio isotopes (literature references) 59
Lead Coatings
  anodic solution 80
  chemical solution 79
  coulometric method, electrolyte for 22
  jet test solution for 15
  guide to destructive tests 86
  guide to non-destructive tests 92
  specifications 111, 116
Lead-Tin Coatings
  guide to destructive tests 86
  guide to non-destructive tests 92
Lea Manufacturing Co.
  Lectromag 31, 102
Lectromag 31, 102
Leitz, E., GmbH
  interference microscope 53, 104
  Tolerator 8, 90
Leptometer 46, 102
Leptoscopes 35, 102
Leybold GmbH
  optical coating thickness meter 106
Liebhafsky, H. and Zemany, P.
  'Film thickness by X-ray emission spectrography' (reference) 66
Light Section Method 51, 111, 112
List, Heinrich
  SM and Diameter coating thickness meters 41, 98, 108
Lorenz, F. R. see Read, H. J.

Magne-Gage 28, 102
Magnetic Amplifier Methods 32

Magnetic Attraction Methods 27, 92, 110, 111, 114, 115, 116, 117
Magnetic Coatings
guide to non-destructive tests 92
Magnetic Induction Methods 32, 93, 94
Magnetic Permeability
difference required for measurement by eddy current method 42
Magnus Junior 32, 104
Mahr, Carl
Millimess 8, 90
Makiola, C.
'Coating thickness determination methods' (reference) 26
MA S/P Magnetic Thickness Gauge 31 104
Material Combinations
ease of measurement 26
ease of measurement by eddy current methods 42
Mathur, P. B. and Lakshmanan, A. S.
'Chemical methods of testing the thickness of electrodeposits' (reference) 11
Measurement and Control System FH 46 59, 100
Mechanical Methods of Test 1, 27
Mesle Chord Method 7
Meters 2, 7, 27
Methods see Destructive Tests; Non-Destructive Tests; and the individual methods.
Methylene Blue Titration
for quantitative determination of tin 11
Meuthen, B.
'The coulometric determination of coating thickness of metallic deposits on steel plate and tubes' (reference) 17
Michelson Principle
applied to set-up of two-beam interference microscopes 53
Micro-Derm 62, 104
Microscopes 1, 52, 93, 94, 95, 102, 104, 106, 111, 112, 114, 115, 116
Microscopes Nachet
interference microscopes 53
Mikrotest 28, 104
Millimess 8, 90
Mini-Leptoscope 104
Minitector 37, 104
Minitor Thickness Gauge 46, 104
Monimeter 36, 106
Müller, P.
'Methods and instruments for non-destructive coating thickness measurement' (reference) 26
Multiple-Beam Interferometry 54

Nachet 300 Microscope 106
Names and Addresses 118
Nash and Thompson Ltd. 50
Netherlands Specifications 114
Nickel Coatings
anodic solution 80
calorimetric method for 16
chemical solution of 73
coulometric method, electrolyte for 22
different bright, giving different thermoelectric effects 50
guide to non-destructive tests 92
jet test solution for 13
specifications 110, 113, 116, 117
Nickel-Copper-Nickel Coatings
chemical solution 73
jet test solution for 13
specification 117
Nickelscope 44, 106
Non-Destructive Tests
principles and general notes 26
general specifications 110
Methods:
capacitative 49
crystal oscillation 68
dial gauge and weighing 27
eddy current 42
electrical resistance 68
magnetic attraction 27
magnetic induction 32
optical 51
photoelectric 55
radiation 56
thermoelectric 50
X-ray 66
Equipments:
Accuderm 38
Atomat 58, 60, 67
Balzers Quartz Oscillator Thickness Monitor 68
Beta 750 Gauge 60

Non-destructive test equipments (continued)
Betameter 60
Betascope 60
BNF thermo-electric thickness meter 50
BSA-Tinsley Pencil Gauge 29
Burndept 60
Caviderm 68
Chemigage 29
Dektak 10, 88
Dermitron 42
Diameter 41
Edwards film thickness monitor 68
Edwards optical film monitor 56
Elcometer eddy gauge 47
Elcometer model 101 32
Elcotector 47
Elektrotest 39
Elmicron 38, 48
Filmeter 47
Fluoroscopy Gauge 66
Frieseke and Hoepfner System FH 46 59
G.E.C. reluctance-type 41
Hacker interference microscopes 55
Inspector 28
Isoscope 44, 102
Isometer 45
Lectromag 31
Leitz intereference microscope 53
Leitz light section microscope 52
Leptometer 46
Leptoscope 35
MA S/P 30
Magne-Gage 28
Magnus Junior 32
Micro-Derm 60
Mikrotest 28
Minitor 46
Minitector 37
Monimeter 36
Nickelscope 45
Norelo meters 67
Permascope 38, 44
Philips X-ray fluorescence gauges 67
Pull-Off Gauge 29
Schaefer Quartz Oscillator Thickness Monitor 68
Sloan Instrument Deposit Thickness Monitor 68
SM coating thickness meters 41
Ultrasonoscope 48
Watson interference microscope 53
Zeiss interference microscope 53
Zeiss light section microscope 52
Non-Magnetic Coatings
guide to non-destructive tests 92
Norelco
surface density meters 67, 106
Norwegian Specifications 114
Nuclear Enterprises Ltd.
Atomat gauges 58, 60, 67, 96

Optical Coating Thickness Meter (Leybold) 106
Optical Film Monitor 56, 106
Optical Methods
guide to use of 93, 94, 95
interference methods 53
light section method 51
photoelectric method 55
specifications 111, 112, 114, 115, 116
Organic Coatings
dry 4
guide to destructive tests 84
guide to non-destructive tests 92, 94
metallised 65
thickness determination specifications 113, 115, 117
transparent 52
wet 3
Oscillating Crystal Methods 68
Oxide Films
chemical solution 71

Paint Coatings (see also Lacquer Coatings)
dry 4
guide to destructive tests 84
guide to non-destructive tests 92, 94
metallised 65
wet 3
specifications 113, 115, 117
Paint Inspection Gauge 6, 90
Panax Equipment Ltd.
Beta 750 Plating Thickness Gauge 64, 98

Pantzke, D. see Elze, J.
PEG Thickness Gage 5
Pencil Thickness Gauge 29, 98
Penetrometer 4
Permascope 38, 44, 106
Permeability
  difference required for measurement by eddy current method 42
Perthen, Dr.-Ing. GmbH
  Elmymeter 88
Pfund Film Thickness Meter 4, 90
Philips Electronic Instruments
  Norelco zinc/aluminium and tin surface density meters 67, 106
Philips Industrie Elektronik GmbH
  PW 4390 Coating Thickness Meter 67, 108
Phosphate Coatings
  chemical solution 73, 75
  specification 114
Photoelectric Method 55
Plastic Foils
  thickness measurement 58
Platimeter 8, 90
Platinum Coatings
  by chemical dissolution of basis metal 12
  by radiation method 65
  guide to destructive tests 86
  guide to non-destructive tests 92
Polish Specifications 115
Potential-Time Diagrams 18
Preece Immersion Test 15
Principles
  destructive methods 1
  non-destructive methods 26
Printed Circuit Through Holes 63,68,97
Pulker, H. and Ritter, E.
  'Short review of methods of determination of thickness of thin layers' (literature reference) 108
Pulker, H. K.
  'Investigation of continuous thickness measurement of thin vapour-deposited layers with a quartz oscillator measuring equipment' (reference) 108
Pull-Off Gauge (Elcometer) 30, 98
PW 4390 Coating Thickness Meter (Philips Industrie) 67, 108

Quartz Oscillator Thickness Monitors (Schaefer GmbH and Balzers A.-G.) 68, 108

Radiation Methods 56, 92, 93, 94, 116
Radiation Transmission Method 57, 116
Radioactive Sources 57, 66
Rapp, T., KG
  Platimeter 8, 90
Reaction, Heat of, Method 16
Read, H. J. and Lorenz, F. R.
  'Comparison of methods for nickel and acid copper on steel' (reference) 13
Reluctance-Type Thickness Gauge 41, 108
Resistance (Electrical)-Measuring Instruments 68, 116
Rhodium Coatings
  by radiation method 65
Riccio, V. A.
  'A non-destructive method for the measurement of the individual coatings thickness in the copper and nickel electrolytic deposits on ferrous base' (references) 43
Ritter, E. see Pulker, H.
Rossmann Dry Film Thickness Gauge type 233 9, 90
Rossmann Wet and Dry Film Thickness Gauge, Type 296 9, 90
Rossmann Wet Film Thickness Gauge 3, 90
Ruder, R.
  'Coating thickness determinations of liquid and pasty materials on substrates' (literature reference) 2
Rumanian Specifications 115
Rupp, H. see Fischer, H.
Russian Specifications 117

S1566 and S1567 Coating Thickness Meter (Kröplin) 9, 88
Schaefer GmbH
  quartz oscillator thickness monitor 68, 108
Schneider, H. see Diesing, P.
Screws
  specification 113

Sectioning Methods 1, 7, 110, 111, 113, 114, 116, 117
Seddon Test 19
Sel-Rex Ltd. see EFCO Ltd.
Semi-Conductor Method of Non-Destructive Thickness Measurement 55
Sheen Instruments Ltd.
wet coating thickness gauge 90
Sherardised Coatings
test specification 113
Silicon
measurement of oxide layer on 55
Silver Coatings
anodic solution of 80
calorimetric method for 16
chemical solution of 74
coulometric method, electrolyte for 22
guide to destructive tests for 86
guide to non-destructive tests for 92
jet test solution for 14
radiation methods for 65
specifications 113, 114
Sloan Instruments Division
Angstrometer 96
Caviderm 68
Deposit Thickness Monitor 68, 98
SM Coating Thickness Meters (List) 41, 108
Solution Methods of Test see Chemical Stripping Methods; Electrolytic Solution Methods
Spark Spectrographs 24
Specifications 110
Spectrographic Methods 24
Spot Tests 15, 111, 112, 115, 116, 117
Spray Metallised Coatings
thickness determination specification 111
Springer, R.
'The removal of coatings' (literature reference) 70
Stierand Prüfgeräte GmbH
Paint Inspection Gauge 90
Straschill, M.
'Modern practice in the pickling of metals' (reference) 70
Strip see Galvanised Wire and Strip
Stripping
basis metal 12
coatings by chemical methods see Chemical Stripping Methods
coatings by electrolytic methods see Electrolytic Solution Methods
Strip Plating Lines
automatic thickness measurement and control in 67
Swedish Specifications 115
Swiss Specifications 116

Tables
chemical solution of coatings 70
destructive thickness measuring instruments 88
electrolytic solution of coatings 76
guide to destructive tests 84
guide to non-destructive tests 92
names and addresses 118
non-destructive coating thickness measuring instruments 96
standard specifications 110
Tantalum
measurement of oxide layer on 55
Tests see Destructive Tests; Non-Destructive Tests; and the individual tests.
Thermoelectric Method 50, 116
Thickness Gauge Model 101 (Elcometer) 33, 100
Thickness Tests see Destructive Tests; Non-Destructive Tests; and the individual thickness tests.
Thorn Bendix Ltd.
coulometric plating gauge 21, 88
thermo-electric thickness meter 50
Threads
specification 113
Through Hole Plating 63, 68, 97
Tin Coatings
anodic solution of 82
by Bendix (anodic stripping) method 23
by coulometric method, electrolyte for 22
by potential-time diagram 18
by radiation method 65
by Seddon test 19

chemical solution of 74, 112
guide to destructive tests for 87
guide to non-destructive tests for 92
jet test solution for 14
on brass and copper 11
specifications 110, 112, 113, 115, 116, 117
strip line control by X-ray fluorescence 67

Tin-Nickel Coatings
specification 113

Tin-Zinc Coatings
guide to destructive tests 87
guide to non-destructive tests 92
jet test solution for 14

Titanium Coatings
guide to non-destructive tests 94
on paper or plastic, control by X-ray fluorescence 67

Tolansky, S.
'Multiple-beam interferometry' (reference) 54

Tolerator 8, 90

Transformer Methods 32

Twin City Testing Corp.
Betascope 60, 96

U.K. Specifications 113

Ultrasonoscope Film Thickness Meter 48, 108

Unit Process Assemblies Inc.
Accuderm 38, 96
Dermitron 43, 98
Micro-Derm 62, 104

U.S.A. Specifications 117

U.S.S.R. Specifications 117

Vacuum Deposits 55, 95

Varnishes see Paint Coatings

Watson, W. and Sons Ltd.
interference microscope 53

Wax Foils
thickness measurement 58

Weiner, R.
'Coating thickness measurement of electrolytic deposits' (reference) 70

West German Specifications 112

Wet and/or Dry Film Meters or Gauges 2, 27

Wet Paint Coatings
thickness determination specification 117

White, R. A.
'Coulometric plating thickness meter' (literature reference) 17

Wire
tin coated (see also Tin Coatings) 19
zinc coated (see also Zinc Coatings) 11, 18, 67, 112, 113

X-Ray Methods 57, 66

Zaba Apparatus 11, 90

Zeiss, Carl
interference microscope 53, 104
light section microscope 52, 104

Zemany, P. see Liebhafsky, H.

Zinc Coatings
anodic solution of 82, 112
by potential-time diagrams 18
calorimetric method for 16
chemical solution of 75, 112
coulometric method, electrolyte for 22
guide to destructive tests for 87
guide to non-destructive tests for 92
jet test solution for 14
on wire 11
specifications 110, 111, 113, 116, 117
strip line control by X-ray fluorescence 67

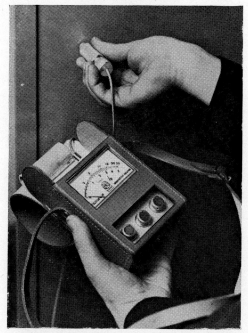

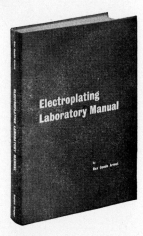

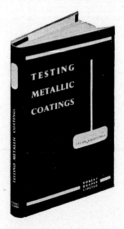